# Biofilm Cities

*How Microbes Build, Defend, and Communicate*

Team Trex

Published by NobleTrex Press



For permissions and other inquiries, write to:
P.O. Box 3132, Framingham, MA 01701,
USA

# Contents

# Introduction

Imagine a sprawling metropolis. It boasts a complex skyline of high-rise towers connected by a network of supply channels that function like superhighways, delivering food and carrying away waste. It has a distinct division of labor, a centralized communication system that pulses with chemical chatter, and a formidable defense strategy that can withstand assaults from the most advanced weaponry known to science. Its citizens are cooperative, specialized, and fiercely protective of their home.

Now, shrink that image until it fits comfortably on the tip of a needle.

This isn't science fiction. It is the reality occurring on the rocks in your local river, inside the pipes under your sink, on the hull of the nearest cargo ship, and, most intimately, on the surface of your teeth at this very moment. We tend to think of cities as a uniquely human invention-a testament to our ability to organize, engineer, and dominate our environment. But we are late to the party. Bacteria have been building cities for billions of years. We call them biofilms.

For much of the history of science, we misunderstood the microbe. If you recall your high school biology classes, you likely picture bacteria as solitary drifters-lonely, microscopic swimmers floating freely in a test tube or a drop of water. This state, known as "planktonic," is how we have studied them for over a century. We grew them in nutrient-rich broths where they swam apart from one another, treating them as fiercely independent individuals. We built our entire understanding of microbiology, and our entire arsenal of antibiotics, based on the behavior of these solitary wanderers.

There was just one problem: in the real world, the solitary wanderer is a myth.

In nature, bacteria almost never choose to be alone. Given the slightest chance, a free-floating microbe will latch onto a surface-a stone, a leaf, a medical implant, or human tissue-and undergo a radical transformation. It sheds its ability to swim and begins to excrete a sticky, protective slime. It signals its brethren to join it. Together, they construct a fortress. This was the paradigm shift championed by the late Bill Costerton, often called the "father of biofilmology," who forced the scientific community to accept that the "pure culture" of the laboratory was an artifact, not reality. He showed us that the natural state of bacteria is social.

This book is the story of that social life. It is an exploration of the invisible metropolises that cover the planet. When we look through this lens, the microbial world stops looking like a chaotic soup of germs and starts looking like a civilization.

The realization that microbes build cities changes everything about how we understand life on Earth. It explains why a pristine mountain stream can clean itself, yet a hip implant can develop an infection that no pill can cure. It explains how the titanic structures of offshore oil rigs are slowly eaten away by microscopic rust-feeders. It challenges our definitions of individuality and intelligence. If a million single-celled organisms can work together to build a water filtration system, coordinate a defense against a predator, and share genetic "updates" like software patches, are they really just simple cells? Or are they something more-a super-organism, a collective intelligence operating on a scale we are only just beginning to comprehend?

Our exploration begins where the science of microbiology itself began: in the human mouth. When Antonie van Leeuwenhoek first turned his handcrafted lenses toward the scrapings of his own teeth in the 17th century, he didn't just find "animalcules"; he found a fully functioning ecosystem. The dental plaque that we scrub away every morning is the archetype of the biofilm city. It is a dense, multi-species jungle where different bacteria occupy specific neighborhoods, trading resources and fighting for territory.

But how does such a complex community arise from nothing? The process of building a biofilm is as dramatic as the founding of Rome. It starts with the "landing," a precarious moment where a single cell must overcome the physics of flow and repulsion to gain a foothold. We will trace the mechanics of this first handshake between microbe and surface, seeing how they deploy specialized grappling hooks to anchor themselves against the current.

Once the foundation is laid, the construction boom begins. The citizens of these cities do not simply pile up; they secrete a substance known as the Extracellular Polymeric Substance, or EPS. This is the "dark matter" of the biofilm, a matrix of sugars, proteins, and DNA that acts as the city's concrete, its glue, and its shield. As we delve into the architecture of these communities, we will find that the slime is not unstructured goo. It is engineered. Mature biofilms form mushroom-shaped towers and intricate canyons designed to maximize nutrient flow. They create oxygen gradients that allow anaerobic bacteria-those that die in the presence of air-to survive safely in the deep basement levels of the city, shielded by their oxygen-breathing neighbors in the penthouse suites.

Perhaps the most unsettling discovery in modern microbiology is that these citizens talk to one another. For a long time, we assumed bacteria were deaf and mute. We now know they possess a molecular language, a phenomenon known as quorum sensing. They constantly release chemical signals into the environment-a "census" of sorts. When the population density reaches a specific threshold-a quorum-the concentration of these chemicals triggers a simultaneous genetic shift across the entire community. Suddenly, millions of individual bacteria stop acting as individuals and begin acting as a group. They might simultaneously release a toxin to attack a host, or produce a glowing light to mimic a squid's camouflage. We will decode this "Esperanto of the microbes," understanding how they count their neighbors and coordinate attacks with military precision.

However, where there is society, there is politics, and biofilm cities are rife with political drama. Evolution dictates that individuals act in their own self-interest, yet biofilms run on cooperation. Bacteria produce expensive enzymes to digest food that benefits their neighbors. They manufacture the protective matrix that shields the whole group. This creates a vulnerability: the cheater. We will meet the "freeloader" mutants that stop contributing to the public good, draining the city's resources while giving nothing back. The way biofilms police their communities-identifying and starving out these cheaters-offers a fascinating mirror to our own social dilemmas regarding public goods and altruism.

The stakes of understanding these cities could not be higher. In the realm of human health, the biofilm is a formidable enemy. If you have ever wondered why a simple urinary tract infection can become a recurring nightmare, or why the lungs of Cystic Fibrosis patients become a battleground that antibiotics cannot clear, the answer lies in the fortress mentality of the biofilm.

We will clarify a critical distinction that often confuses even the medically savvy: the difference between resistance and tolerance. Resistance is genetic; it's a mutation that renders a drug ineffective. Tolerance, however, is the superpower of the biofilm. When an antibiotic wave washes over a biofilm city, the outer layers may perish, neutralizing the drug in the process. But deep inside, shielded by the matrix and deprived of nutrients, reside the "persister cells." These are the sleeper agents of the bacterial world. They enter a dormant, zombie-like state where they are metabolically inactive. Since most antibiotics work by sabotaging active cellular machinery, they pass harmlessly over these sleeping cells. Once the antibiotic treatment stops, the persisters wake up and rebuild the city from the ashes. It is a strategy of patience that frustrates our best medical technologies.

This resilience extends beyond the human body. In the industrial world, biofilms are agents of entropy. We will investigate the silent catastrophe of biofouling, where microbial cities colonize the hulls of ships, creating drag that costs the global economy billions in fuel. We will look at the phenomenon of biocorrosion, where microbes literally eat the infrastructure of the modern world, dissolving oil pipelines and weakening bridges from the inside out.

Yet, we must be careful not to paint these cities solely as villains. We are, in a very real sense, walking biofilms ourselves. Our gut microbiome is a massive, productive city that digests our food and trains our immune system. Furthermore, we are learning to harness the power of these communities. In wastewater treatment plants, we rely on massive biofilm cities to break down our sewage, turning toxic sludge into clean water. We are finding ways to recruit "good" citizens to displace the bad ones, using ecological competition rather than chemical warfare.

As we move toward the final chapters of this story, we look to the future. How do we dismantle a city that was built to survive the apocalypse? Scientists are currently developing a new generation of tactics that move beyond the brute force of antibiotics. We are learning to jam their communication signals, effectively blinding the bacteria to their own population numbers. We are engineering enzymes that dissolve the EPS mortar, turning the fortress back into a vulnerable crowd of individuals. We are even studying the natural "dispersal" events where biofilms spontaneously decide to break camp and leave, hoping to trick infections into fleeing the body of their own accord.

The journey ahead will take us from the microscopic geometry of a slime mold to the sterile corridors of a hospital ward, and from the deep-sea vents where life may have originated to the shiny surfaces of the International Space Station, where biofilms have already established a foothold.

To understand the microbe, we must stop looking at the cell and start looking at the community. We must stop thinking of them as simple germs and start respecting them as the original architects of Earth. They built the first cities, they invented the first communication networks, and they perfected the art of defensive warfare billions of years before we sharpened our first stick.

Welcome to the metropolis beneath the microscope. Welcome to the biofilm city.

Chapter 1

# The Invisible Metropolises

*Under our feet and within our bodies, vast empires rise and fall without us ever noticing. This chapter peels back the layer of the invisible to reveal that bacteria are not merely drifting loners, but sophisticated city-builders capable of constructing fortresses that rival human engineering.*

## 1.1 A Zoo in the Mouth

If you want to meet your oldest, most faithful companions, don't look in your photo albums. Run your tongue along the back of your teeth. That faint roughness, the thin film that returns a few hours after brushing, is not food debris in any simple sense. It is a living settlement. Every time a tooth feels slightly fuzzy, a city is forming.

The mouth is a tempting place to imagine as clean. It is bright, easy to rinse, constantly bathed in saliva. Yet from a microbe's point of view it is prime real estate: warm, wet, regularly fed, and stitched with surfaces that do not shed as quickly as skin. The cheeks and tongue are like moving carpets, constantly rubbed and renewed. Teeth are more like cliffs. They stand still, hard as stone, offering microbes a permanent address. Add to that a daily schedule of sugar, starch, proteins, and fats, and you have a habitat that rewards those who can attach, cooperate, and endure.

The first person we know to watch this drama with any clarity did so in the late seventeenth century, by candlelight and curiosity. Antonie van Leeuwenhoek, a Dutch draper turned microscopist, built tiny lenses of startling quality and aimed them at almost anything he could scrape, soak, or suspend in water. He was not trained in medicine. That was part of the advantage. He carried none of the assumptions that later taught students would inherit about what living things were supposed to look like. He simply looked, and described what he saw with the careful stubbornness of an artisan.

One day, he took material from his own mouth: the whitish matter from between his teeth. It is an intimate detail that still has the power to make a reader recoil, and that recoil is useful. It reminds us that scientific discovery often begins in places too ordinary to respect. Van Leeuwenhoek mixed the scrapings with water and placed the drop before his lens. What he saw was not a quiet sediment. It was motion: countless small creatures, darting and wriggling, some moving like eels, others tumbling as if pushed by invisible currents. He called them *animalcules*, little animals.

There is a psychological jolt in that moment. Dental plaque sounds inert, even trivial; *animalcules* sounds like a menagerie. Van Leeuwenhoek's word did not merely label a new object; it dragged microbial life into the category of the animate. These were not stains or vapors or a chemical haze. They had shapes and habits and, above all, agency. They moved with *vigor*. If you had grown up in a world where the smallest living creatures were fleas and lice, the idea that your teeth sheltered a zoo would have felt like a private apocalypse.

Van Leeuwenhoek was careful to compare mouths. He reported that he found fewer *animalcules* when he scraped after cleaning his teeth, and more when he did not. He compared his own mouth with those of others, including people with poorer oral hygiene. He did something else, almost casually, that reads today as a proto-ecological observation: he noticed that the same kind of scraping could contain different kinds of creatures. Even without the vocabulary of species, communities, or ecosystems, he had stumbled into the central truth of oral life: the mouth is not a single habitat but many, and the inhabitants vary from one micro-neighborhood to another.

It would be easy to treat this as a charming prologue to modern microbiology, a historical curiosity before the real science begins. That would miss what is most unsettling about it. Van Leeuwenhoek did not merely discover *microbes*. He discovered density. The mouth was not host to a few occasional intruders. It was packed. A smear from a tooth could contain a multitude so thick that the water seemed to shimmer with movement. The striking thing was not that life existed at small scales; it was that so much life could occupy so little space.

A useful way to picture this is to translate plaque into the language we reserve for crowds. The surface of a tooth is small enough to ignore, but microbes do not live on the scale of teeth. They live on the scale of micrometres, and on that scale a tooth is a continent. A community can stack itself in layers, build shelters, open channels for flow, and create protected pockets where the chemistry is different from the saliva outside. A human city is impressive because it concentrates bodies, food, waste, and information in a tight space. Plaque does the same, but with cells a thousand times smaller and with architectures that are almost impossible to appreciate until you have learned to stop expecting neatness.

The mouth offers another kind of misdirection. Because saliva is always moving, it feels as if everything should be washed away. Yet the most successful oral microbes are not the drifters. They are the settlers. The first act in this story is attachment: a bacterium makes contact with a surface and stays. Teeth, newly cleaned, do not remain bare for long. Within minutes they are coated by a thin film derived from saliva, sometimes called a pellicle. It is not dirt; it is a natural layer of proteins and other molecules that immediately adsorb onto enamel. To microbes, it is not a protective varnish so much as a landing pad.

Once a few cells attach, the settlement becomes easier for others. Early colonizers alter the surface, provide binding sites, and change the local conditions. Soon the community begins to produce an extracellular matrix, a sticky material that is part scaffold, part glue, part sponge. This matrix is often described casually as slime, but that word suggests something accidental and formless. In reality it behaves more like infrastructure. It can hold cells in place against the shear forces of saliva and chewing. It can trap nutrients and keep them nearby. It can slow diffusion, meaning that acids and other chemicals do not instantly disperse. It can even create microenvironments, small chemical worlds where oxygen is scarce or the pH is lower than in the surrounding saliva.

The result is a paradox that van Leeuwenhoek could not have guessed from motion alone: the most dynamic lives in the mouth often occur inside communities that are physically anchored. The zoo is not a swarm in open water. It is a city built on rock.

Thinking of plaque as a city also helps explain why it can be both ordinary and dangerous. Cities are not inherently malignant. They concentrate resources and allow collaboration; they also concentrate waste, conflict, and vulnerability. Dental plaque is similar. In many mouths, much of the time, plaque behaves less like an invading army and more like a managed border community. The resident microbes occupy space and consume available nutrients in ways that can discourage newcomers from establishing themselves. The details are complex, but the basic idea is familiar: a crowded, stable neighborhood leaves fewer empty apartments for strangers.

Trouble begins when the balance of the neighborhood shifts. The mouth is fed in pulses. A sip of sugary coffee, a bite of bread, a snack eaten at a desk: these are sudden deliveries of fuel, and microbes are quick to exploit them. Many bacteria can ferment sugars and produce acids as waste products. If those acid pulses are frequent, the local environment near the tooth surface can become repeatedly acidic. That matters because enamel is not invincible; it is a mineral that can dissolve under acidic conditions. It also matters because the community itself changes under pressure. Some microbes tolerate acidity better than others, and repeated low-pH episodes select for those acid-tolerant residents. Over time the community can tilt toward a more acid-producing, acid-enduring population. Cavities are not simply the work of a single villainous species; they are a failure of ecological diplomacy, a shift in the city's economy toward an industry whose exhaust corrodes the buildings.

The same ecological logic applies to the gumline. The crevice between tooth and gum is a peculiar habitat, sheltered from the full force of saliva and constantly influenced by the immune system. Inflammation changes the chemistry of that crevice, altering what nutrients are available and how much oxygen penetrates. These changes can favor different kinds of microbes, often those that thrive in low-oxygen conditions and can make use of the proteins and other molecules that become more abundant during inflammation. Here again, disease can look less like a simple invasion and more like a change in zoning laws: the neighborhood's conditions shift, and a different set of residents finds it profitable to expand.

None of this should make you sentimental about plaque. The point is not that microbial cities are benevolent. The point is that they are *organized*. The mouth is not merely contaminated. It is colonized, structured, and negotiated.

Van Leeuwenhoek's observation also raises a question that still has the power to unsettle: how can so many distinct lives coexist in such close quarters? Part of the answer lies in the way the mouth is divided into habitats. The tongue is a rough, muscular landscape. The cheeks are smoother and constantly rubbed. The roof of the mouth is relatively stable. Teeth are hard, non-shedding islands. Even within a single tooth, the front-facing side and the tongue-facing side experience different flows of saliva, different abrasion, and different exposure to food. Tiny differences in surface texture and fluid movement can create different opportunities for attachment and growth. If you imagine the mouth as one room, these differences are easy to miss. If you imagine it as an archipelago, each island with its own climate, diversity becomes less mysterious.

Another part of the answer lies in cooperation and conflict at microscopic scales. In a crowded biofilm, one microbe's waste can become another's food. Some consume oxygen quickly, creating low-oxygen zones where strictly anaerobic organisms can thrive. Some produce substances that inhibit competitors. Some benefit from the enzymes and matrix produced by others, living as opportunists within a public works project they did not pay for. A plaque biofilm can behave like a layered economy: producers, scavengers, specialists, and cheaters all entangled in physical proximity. The most counterintuitive feature is that this complexity can be stable. A community can return to its familiar composition after disturbances, not because it is rigid, but because its network of interactions tends to rebuild itself in predictable ways.

If van Leeuwenhoek had been able to freeze time and slice through plaque as one slices through a cake, he might have been most surprised by its architecture. Biofilms are not uniform carpets. They can have towers and channels, dense clusters and looser regions.

Water can flow through microscopic channels, delivering nutrients and removing waste. Cells can sit in pockets where conditions differ sharply from those just a few micrometres away. To a human eye, plaque seems like a smear. To a microbe, it is a landscape.

The idea of a landscape brings us back to a more intimate truth: brushing and flossing are forms of urban planning. They are not merely attempts to sterilize. Complete sterilization is impossible, and arguably undesirable. The mouth is re-seeded constantly from saliva, from food, from other people, from the throat. What brushing does, at its best, is disrupt the physical structure of the biofilm before it becomes mature and entrenched. It breaks up the matrix and removes the organized communities that make microbial life resilient. The goal is not to remove every cell; it is to prevent the formation of thick, protected neighborhoods where acids can pool and where oxygen gradients can shelter more problematic residents.

This makes an odd kind of sense of the feeling of “clean”. After brushing, teeth are smooth because the biofilm has been physically reduced; the cliff face has been scraped. But smoothness is temporary because surfaces invite settlers. Within hours, early colonizers return, and the pellicle reforms. Cleanliness in the mouth is therefore not a state achieved once but a rhythm maintained, a repeated interruption of city-building.

There is a temptation, when hearing about microbial density and complexity, to imagine oneself as outnumbered and therefore doomed. But coexistence is older than anxiety. The mouth has been home to microbial communities for as long as there have been mouths. The immune system is not a simple exterminator; it is also a manager, tolerating many residents while restraining those that cause damage. Saliva is not just water; it contains proteins that bind microbes, enzymes that can degrade molecules, buffering systems that resist acidification, and constant flow that makes attachment a challenge. Teeth and gums are not passive substrates; they provide surfaces, but they also impose constraints, from mineral chemistry to immune surveillance at the gumline.

The discovery of *animalcules* did something else, too: it destabilized the old boundary between the inside and the outside. If creatures could live on your teeth, then you carried wilderness within you. That idea has echoes in modern discomfort about germs, but it can also provoke a different reaction: awe. The mouth is not simply a portal through which the world enters the body. It is a habitat in its own right, with a resident population that responds to diet, hygiene, stress, and health. It is both intimate and foreign, familiar and alien.

It is worth lingering on why van Leeuwenhoek's *animalcules* were so lively. Motion, in his lens, was evidence of vitality, and he was right. Yet motion also misled. Many bacteria in mature plaque are not swimming freely. They are embedded in a matrix, held in place, living in close contact. Under the microscope, the most dramatic movement may come from organisms that are not the dominant architects of plaque but rather the ones capable of darting through liquid when given the chance. In other words, the most visible citizens are not always the most influential. The quieter majority may be the builders: the cells that attach early, produce matrix, and create the conditions that determine who else can thrive.

That mismatch between visibility and importance is a recurring theme in microbiology. Humans notice what moves, what smells, what discolors, what causes pain. Microbial life often matters most when it becomes structured: when it forms layers, changes chemistry, and persists. Dental plaque is the daily demonstration. It is not impressive because it is rare; it is impressive because it is routine.

A final, concrete thought experiment brings the scale into focus. Imagine leaving a bowl of sugary tea on your kitchen counter for a few days. You might see a film form, perhaps a faint skin. That is not so different in principle from what happens on teeth after meals, except that teeth are not a neglected counter: they are constantly rinsed, buffeted, and defended. And still the film returns. The success of plaque is not a sign that the mouth is dirty; it is a sign that microbes are good at making homes.

Van Leeuwenhoek's dental scrapings were, in the end, not merely a glimpse of unseen creatures. They were a portrait of community life at a scale the human brain did not evolve to imagine. The mouth, so closely tied to speech, taste, and intimacy, turns out to be one of the most densely inhabited places most of us will ever encounter. A city can be built without stonecutters, without architects, without anyone who knows the whole plan. It can be built by countless small lives following local rules: attach here, feed there, excrete this, avoid that, cling together when the currents rise. And each time you close your jaw, you carry that city with you, a private metropolis rebuilt again and again on the enamel cliffs.

## 1.2 The Planktonic Illusion

A curious thing happens when a science is born in glassware: the container begins to feel like the natural world.

For much of the twentieth century, the default picture of a bacterium was a solitary cell drifting in a clear broth, multiplying until the liquid turned cloudy. That image was not foolish. It was powerful, practical, and, in many ways, lifesaving. If you want to show that a microbe causes disease, you need to isolate it. If you want to test whether a drug can stop it, you need a standard way to grow it and count it. A test tube full of liquid is easy to inoculate, easy to mix, easy to sample, and easy to measure. From a distance, a culture flask offers what the messy world rarely provides: control.

The trouble is that control can become a costume for reality.

Microbiologists borrowed a word from oceanography: *planktonic*. Plankton are organisms that drift, carried by currents rather than anchored to a surface. A *planktonic bacterium* is, by analogy, a free-floating cell, suspended in liquid and living more or less on its own. In the laboratory, planktonic life is convenient. Bacteria distribute through the medium; nutrients are shared; oxygen can be introduced by shaking; growth can be tracked by optical density, a quick measurement of cloudiness that correlates with how many cells are present. A whole culture can be treated as if it were a uniform soup.

Outside the laboratory, bacteria often refuse to behave like soup.

A more faithful word for how microbes commonly live is *biofilm*. A biofilm is a community of microorganisms attached to a surface and embedded in a self-produced matrix: a hydrated, sticky scaffold built from large molecules released by the cells. That matrix is sometimes called *extracellular polymeric substances* (often shortened to EPS), a dry term for something that behaves like living architecture. The cells are not merely stuck to a surface; they are held together, surrounded, and partly protected by a material they collectively make. In a biofilm, the neighborhood matters. Who your neighbors are, what they secrete, how tightly packed you are, and how easily molecules can flow in and out can determine whether you thrive, cooperate, or die.

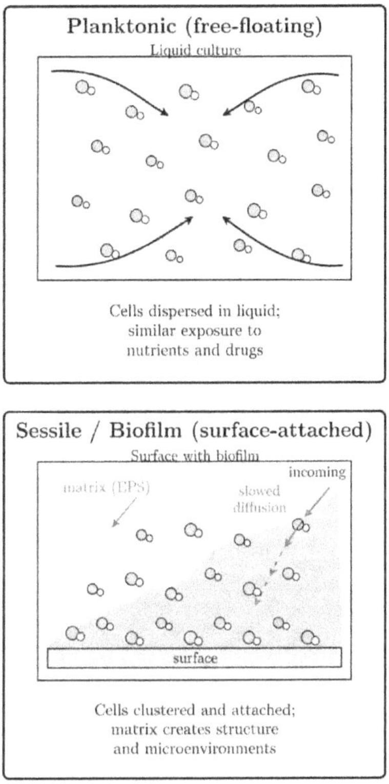


Once that definition is on the table, the old laboratory picture starts to look like a carefully chosen special case. The planktonic cell is real, but it is not always the main character.

Why did the planktonic view dominate for so long? Partly because it fit the tools. A free-floating culture can be treated as a single population: well-mixed, uniform, and easy to count. A biofilm is awkward. It is attached. It is lumpy. It refuses to distribute evenly through the liquid. If you scrape it from a surface, you break it apart, and the act of measuring changes the thing you want to measure. Much of microbiology's early success depended on making organisms behave in standard, repeatable ways. Biofilms are repeatable too, but they require different habits of thought: surfaces, flow, structure, and spatial gradients rather than a homogeneous broth.

There was also a strong conceptual current in medicine that reinforced the planktonic illusion. The germ theory of disease, crystallized in the late nineteenth century, taught clinicians and scientists to look for a specific culprit that could be isolated, cultured, and linked to a particular illness. This framework was enormously fruitful. It led to vaccines, sanitation, and antibiotics. Yet it also encouraged a kind of narrative simplicity: one pathogen, one disease, one test tube. Communities of many species, clinging to surfaces and acting together, sit uneasily in that storyline. They are harder to isolate, harder to name, and harder to blame.

A culture flask, in other words, did not merely hold bacteria; it held an idea about what bacteria *are*.

The discovery that biofilms are central to many infections did not arrive as a single dramatic revelation. It arrived as a slow irritation: clinicians saw infections that behaved strangely, engineers saw slime that would not go away, and microbiologists kept encountering results that did not fit their neatest models. Antibiotics could wipe out bacteria in a lab dish, yet the same drug could fail against an infection on a medical device. A swab might show only a few organisms, yet the inflammation persisted. A patient could improve during treatment and relapse soon after. These patterns were often explained away as poor compliance, inadequate dosing, or the emergence of resistance. Sometimes those explanations were correct. But sometimes the deeper issue was that the bacteria were not living as plankton.

One of the pivotal figures in naming and promoting the biofilm concept was J. William Costerton, whose work from the 1970s onward helped make a scattered set of observations cohere into a single idea.

Costerton and colleagues looked closely at bacteria in natural and clinical settings and found a recurring pattern: surface-attached communities surrounded by matrix, showing behaviors very different from the same species grown in liquid culture. The novelty was not the existence of slimy layers; people had long known about plaque on teeth and slime on rocks. The novelty was to treat these layers as a dominant microbial lifestyle with its own biology, not as an annoying residue.

The word *biofilm* itself did a great deal of work. It gave a single handle to phenomena that had been filed in different mental drawers: the stubborn coating inside water pipes, the fouling of ship hulls, the persistent infections of catheters, the plaque on teeth, the chronic lung infections in certain diseases. A name does not create a reality, but it makes a reality easier to see.

Consider a concrete clinical example that has become almost emblematic: a catheter-associated infection. A catheter is a thin tube inserted into the body to deliver fluids or drain urine. It is, from a bacterium's point of view, an invitation: a new surface in a warm, nutrient-rich environment, with flow that can deliver food and remove waste. If bacteria gain access to the catheter surface, they can attach and begin building a biofilm. Once established, that community can become remarkably tolerant to antibiotics and to immune attack. Treatment may reduce free-floating cells in the surrounding fluids, giving the impression of improvement, while the attached community remains as a reservoir. When therapy stops, cells can disperse from the biofilm and seed the infection again.

It is tempting to call that antibiotic failure *resistance*, but that word can mislead. Antibiotic resistance usually refers to inherited genetic changes that allow bacteria to grow in the presence of a drug: enzymes that degrade antibiotics, mutations that change a drug's target, pumps that expel the compound. Biofilm-associated survival often involves something different, sometimes called *tolerance* or *recalcitrance*. Cells deep in a biofilm may grow slowly, making antibiotics that target active growth less effective. The matrix can hinder penetration or bind certain molecules. Chemical gradients can create zones where conditions reduce drug activity. A biofilm can contain a mix of physiological states: some cells active, some dormant, some stressed, some well-fed. Even if none of those cells carry a classic resistance gene, the community can be extraordinarily hard to eradicate.

Laboratory testing quietly reinforced the planktonic illusion here as well. The standard method for gauging antibiotic susceptibility for decades has often involved measuring a minimal inhibitory concentration (MIC): the lowest concentration of a drug that prevents visible growth of a bacterium in liquid culture. MICs are useful, but they answer a particular question: how does a well-mixed population of free-floating cells respond to a drug over a defined time? A biofilm asks a different question: how does a structured, attached community respond when drugs must penetrate a matrix, when growth rates vary by location, and when some cells are effectively hunkered down? A low MIC does not guarantee biofilm clearance, and a clinician can be left with the unsettling gap between what the laboratory predicted and what the patient experiences.

There is a similarly instructive example from industrial settings. Anyone responsible for water systems learns, often with some despair, that bacteria can colonize pipes. The flow that should wash cells away can instead deliver them repeatedly to surfaces. Slight roughness, microscopic cracks, junctions, and gaskets provide protected niches. Once a biofilm forms, it can change the chemistry of the local environment, corrode materials, and release fragments downstream that seed new sites. Engineers may increase disinfectant levels, only to find that the biofilm persists or returns. Again, the issue is not that disinfectants are useless; it is that a surface-attached, matrix-embedded community responds differently than the same organisms would in a beaker.

Biofilms also complicate the very act of *seeing*. A planktonic culture announces itself by turning the liquid cloudy. A biofilm can be nearly invisible: a thin, transparent layer on a surface, a faint sheen in a pipe, a subtle coating on a contact lens. Even when thick, it may appear as a harmless smear rather than a structured community. The sensory cues humans rely on are unreliable at microbial scales. Clean-looking does not mean uncolonized; cloudy does not mean attached.

The illusion persisted, then, because planktonic life is easy to teach and easy to measure. It has a tidy geometry: cells as discrete units, floating in a uniform environment. Biofilms are spatial. They demand that biology be treated almost like geography.

That spatial thinking changes seemingly simple questions. Take oxygen. In a shaken flask, oxygen can be delivered broadly, and the whole culture may be considered "aerobic." On a surface, oxygen can be abundant at the outer edge and scarce a short distance in. A biofilm can contain both oxygen-loving and oxygen-avoiding organisms, layered or clustered according to what diffuses in from the outside. The same goes for pH, nutrients, and waste products. If acids are produced in a dense patch and diffusion is slowed by the matrix, a tiny acidic pocket can form right where enamel is vulnerable. A measurement of saliva pH might look reassuring while, micrometres away, the local chemistry is doing damage.

Even the social lives of microbes look different when the cells are not drifting apart. Many microbial signals are chemical: molecules released into the environment that can be sensed by other cells. In a well-mixed flask, those molecules rapidly disperse. On a surface, trapped within a matrix, signals can accumulate locally. That can make collective behaviors more likely, from the coordinated production of enzymes to changes in gene expression that alter how cells respond to stress. In other words, the matrix does not merely hold cells in place; it can function as a chemical conversation room, keeping messages close enough to matter.

One reason the biofilm idea took time to settle into the mainstream is that it forces a subtle humility about the laboratory. The laboratory is not a neutral window. It is an environment with its own selective pressures. If you repeatedly transfer bacteria by pipetting a small amount of liquid into fresh liquid, you are selecting for organisms that thrive while free-floating. If you choose colonies that grow well on agar plates, you are selecting for those that tolerate that surface, that oxygen level, that nutrient mix. Many organisms that are perfectly successful in nature are difficult to culture under standard conditions, and many interactions that shape community behavior are disrupted when species are isolated from one another. The more a field prizes isolates and pure cultures, the more it risks mistaking a stripped-down performance for the full play.

None of this should be read as an attack on classical microbiology. The planktonic model gave medicine antibiotics and vaccines and gave biology the ability to connect genes to functions in a tractable way. The point is simpler and more unsettling: a model can be both useful and incomplete, and incompleteness matters most when it masquerades as universality.

A good antidote to the planktonic illusion is to pay attention to surfaces. The body is full of them: teeth, gums, the lining of the gut, the airways, the skin, implanted devices, damaged tissue. The nonhuman world is full of them too: rocks in streams, ship hulls, plant roots, the inside of a milk processing line. In all these places, flow and attachment coexist. Life is not only about floating and dividing;

it is about building and staying.

Once you begin to see bacteria as builders, many small puzzles become easier to understand. Why does a freshly cleaned surface become slimy again? Why do chronic infections relapse after apparently appropriate therapy? Why do bacteria in a community behave differently than the same species alone? These are not mysteries of bad luck. They are consequences of a lifestyle in which the group and its scaffolding become part of the organism's effective body.

The planktonic illusion never truly vanishes; it remains a tempting simplification, ready to reassert itself whenever a clear flask turns cloudy. Yet the most consequential microbial worlds often do not announce themselves by cloudiness. They announce themselves by persistence: a film that returns, a smell that lingers, an inflammation that smolders, a pipe that fouls again, a wound that refuses to close. On the surfaces of our lives, microbes are rarely just passing through. They are at home.

## 1.3 The Evolutionary Fortress

A single bacterium is, in principle, a complete life. It can feed, grow, divide, and, in the right conditions, flourish. It can also be killed with casual ease: a sudden dry spell, a change in acidity, a passing dose of detergent, a burst of immune chemistry. The independence of a solitary microbe is real, but it is a precarious kind of freedom, like living outdoors with no roof and no pantry.

So why do bacteria so often give up that independence? Why do they glue themselves to surfaces and to one another, investing energy in a communal scaffold that they could, in theory, have spent on faster growth?

The answer begins with a simple trade: speed for persistence. Free-floating life can be fast when conditions are generous. Surface-attached life can be stubborn when conditions become hostile. The fortress is not a metaphor for aggression so much as for endurance. A biofilm is built to remain when the world tries to sweep it away.

The most visible piece of this endurance is the matrix, the material that surrounds and binds cells in a biofilm. It is easy to imagine the matrix as mucus-like gunk that happens to accumulate, but that picture misses its sophistication. The matrix is assembled from many kinds of large molecules, including polysaccharides (long sugar chains), proteins, lipids, and strands of extracellular DNA. DNA here is not primarily genetic instruction; it is a structural polymer, like rebar in concrete, helping the community hold together. These ingredients can tangle into a hydrated network that is at once sticky and springy. If you have ever tried to wipe plaque off a tooth after it has had time to mature, you have felt a hint of that viscoelasticity: it smears and resists, more like a soft gel than a dusting.

Calling the matrix infrastructure is not poetic license. It changes the physics and chemistry of the microbes' world. It turns a bare surface into a built environment.

Start with the simplest threat: being washed away. In many habitats, from a streambed to the inside of a catheter, there is constant flow. A lone cell that happens to be suspended in that flow can be carried off before it has time to divide. Attachment solves that. A biofilm anchors its inhabitants to a surface and anchors them to each other, so that a gust of fluid does not translate into exile. The matrix spreads the mechanical load; the tug on one cell is shared across many attachments. The community becomes harder to shear away than any one member would be.

Then consider dehydration. Many bacteria are exquisitely vulnerable to drying. The matrix helps by holding water like a sponge. Even when the surrounding environment becomes less hospitable, the interior of a biofilm can remain moist longer, buffering brief dry periods. This is not absolute protection, but it can be the difference between death and survival in habitats that fluctuate.

Predation is another ancient pressure. Bacteria have their own predators: protozoa and other microscopic grazers that engulf and digest cells. A solitary cell is a bite-sized meal. A cell embedded in a matrix is harder to access and sometimes harder to digest. Some predators can graze on biofilms, but the matrix changes the game: it slows feeding, forces predators to work harder, and makes it more likely that the prey survives long enough to reproduce. Even partial protection can be decisive when the timeline of life is measured in minutes.

The most consequential protection, for humans, is against chemical attack. In medicine, that chemical attack often comes in the form of antibiotics and disinfectants, and in the form of immune molecules meant to kill microbes. Biofilms have a reputation for shrugging these off. The reality is more nuanced, and it is precisely that nuance that makes the fortress so effective: there is no single wall to break. There are many overlapping defenses, each imperfect, together powerful.

One layer is physical and chemical interaction with the matrix itself. Some antimicrobial molecules can be slowed as they diffuse through the dense polymer network. Others can be bound or sequestered by matrix components, especially if they carry charges that stick to the polysaccharides or DNA. This does not necessarily stop penetration, but it can delay and dilute the attack. In a fast-growing planktonic culture, minutes matter; in a biofilm, minutes can decide whether a drug reaches deep cells at a lethal concentration or arrives late, weakened, after outer cells have already neutralized some of it.

Another layer is the simple biology of growth. Many antibiotics were designed, consciously or not, to exploit the vulnerabilities of rapid division. Drugs that target cell wall synthesis, for instance, are most effective when a bacterium is actively building and remodeling its wall. If a cell is growing slowly, it presents fewer active targets. Biofilms naturally contain slow growers. They are not necessarily genetically special; they are responding to their position. Deep in the community, nutrients may be scarce, oxygen may be low, waste products may accumulate, and signaling molecules may encourage a different physiological state. The same species can behave like a sprinter on the outside and like a hibernator within.

A third layer is diversity of state. Even in a genetically identical population, cells can occupy different physiological modes: some actively metabolizing, some stressed, some dormant. This heterogeneity is one of the least intuitive features of biofilms for people trained on uniform cultures. In a well-mixed flask, the average behavior can be a good description of most cells. In a biofilm, the average can describe almost nobody. A drug that kills the active majority may leave behind a minority that is temporarily tolerant. When the drug pressure lifts, those survivors can repopulate the structure. Clinicians encounter this as relapse; the microbe experiences it as weather.

The immune system adds another dimension. Immune cells can attack bacteria by engulfing them, by releasing reactive chemicals, and by triggering inflammation that makes the environment hostile. A biofilm can frustrate these strategies. The matrix can act as a barrier to immune cell penetration, forcing the immune response to occur at the surface. The very attempt to attack can injure host tissue, releasing nutrients and creating conditions that favor certain microbes. The result can be a stalemate in which the immune system cannot fully clear the community but continues to respond, producing chronic inflammation. The fortress holds not because the defenders are invincible, but because the battle itself reshapes the terrain.

A concrete medical example makes these abstractions feel less theoretical: infections on implanted devices. A hip replacement, a heart valve, a central venous catheter, even a dental implant offers an unusual surface inside the body. It is durable, non-shedding, and not designed by evolution to be self-cleaning. If bacteria gain access during surgery or later through the bloodstream, they can attach and begin to form a biofilm. Once that community is established, antibiotics that would easily kill the same bacteria in a test tube may fail to eradicate them on the device. The patient can experience cycles of improvement and relapse, or a persistent low-grade infection that never quite resolves. The clinical solution often involves a brutal clarity: remove the device. In fortress terms, it is not always possible to kill the inhabitants without tearing down the walls they live in.

The mouth offers a more familiar example, and a more intimate one, of fortress-building with consequences. Consider what happens after a sugary snack. Many oral bacteria can rapidly metabolize sucrose and other sugars, producing acids. In a thin, loosely organized film, those acids can be buffered by saliva and swept away. In a mature plaque biofilm rich in matrix, diffusion can be slowed. Acids linger. The pH near the enamel surface can drop and remain low for longer, creating a microenvironment that favors acid-tolerant microbes and undermines the mineral stability of enamel. The architecture itself becomes part of the harm: the fortress does not merely protect microbes from the outside; it can trap their own chemical exhaust in precisely the place where it does the most damage.

The same matrix that shelters can also feed. By retaining enzymes and metabolites, the biofilm can behave like an external digestive system. Enzymes secreted by cells do not simply wash away; they can be held close, breaking down complex molecules into smaller, usable parts that remain within reach. Nutrients can be trapped, concentrated, and shared unevenly, creating micro-economies in which some residents benefit more than others. This can look, from the outside, like cooperation, but it is better understood as interdependence shaped by physics. When molecules do not diffuse freely, proximity becomes valuable.

The fortress is not only about defense against enemies. It is also about defense against uncertainty.

Environments change. A stream floods and then dries. A human brushes teeth, eats lunch, sleeps, wakes. A patient takes antibiotics for a week and stops. For a free-floating cell, each change can be catastrophic. For a biofilm, change can be absorbed and localized. Outer layers may bear the brunt of an attack, sacrificing growth or dying, while inner regions remain stable. When conditions improve, surviving cells can expand outward again. This resilience can be selected even if building the matrix slows reproduction in the short term. Evolution does not reward only the fastest divider; it rewards whatever leaves descendants under the conditions that actually occur, including the bad days.

A historical anecdote from the history of antibiotics captures this tension between optimism and persistence. When penicillin and later antibiotics became widely available, there was a brief cultural sense that infectious disease might be solved, that microbes would finally be brought under control by chemistry. That hope was not foolish; antibiotics transformed medicine. But even early on, there were infections that seemed to ignore the new miracle drugs, lingering stubbornly in tissues or returning after treatment. Not all of these cases were due to classic genetic resistance. Some were due to where the bacteria lived and how they organized themselves. The fortress was there all along, hiding in plain sight on teeth, rocks, and medical devices, and it did not crumble simply because a potent molecule existed in the bloodstream.

It is tempting to imagine that a biofilm is a harmonious collective, a community that has decided, as a group, to build walls. Bacteria are not planners. They have no city council. The fortress emerges from local rules executed by individual cells: attach when you encounter a surface that offers stability; secrete polymers when neighbors are near; alter gene expression when signals accumulate; slow down when nutrients are scarce; persist when stress is high. If those rules increase the chance that a lineage survives repeated assaults, they will spread. The impressive structure is an outcome, not an intention.

Even so, something like communication does occur, and it matters for defense. Many bacteria use chemical signaling systems often described as *quorum sensing*: cells release small molecules, and when those molecules reach a threshold concentration, they trigger changes in gene expression. The name can be misleading, suggesting a democratic vote. In reality it is closer to a chemical census, a way to detect whether you are alone in open water or packed into a dense neighborhood where secretions will not be diluted away. In a matrix-rich biofilm, signals can be retained and concentrated, making these systems more effective. The decision to produce certain enzymes, toxins, or additional matrix can be linked to local density and to the presence of other species. Communication, here, is not chatter for its own sake; it is a tool for coordinating expensive behaviors so that they pay off.

The fortress also creates moral confusion for humans because its success blurs the line between resistance and tolerance. When antibiotics fail, the immediate fear is that a superbug has evolved a new gene. Sometimes it has. But sometimes the genes are unchanged and the lifestyle is the difference. The cells are sheltered, slow, spatially arranged, and physiologically diverse. That matters for how we respond. If the problem is genetic resistance, the solution might be a different drug. If the problem is biofilm tolerance, the solution might involve disrupting the matrix, altering surfaces, using combinations of therapies, changing flow, or removing an infected device. The fortress demands tactics, not just stronger bullets.

A less obvious advantage of biofilms is that they can act as archives. In a free-floating population, selection can be ruthless; anything not immediately useful can be lost. In a structured community, pockets of cells can persist in different states, including low-metabolism forms that are less exposed to selective sweeps. This can preserve diversity, which is a kind of long-term insurance. When conditions change, a rare variant or a rare physiological state might suddenly become valuable. The biofilm, by maintaining heterogeneity, can keep options open.

All of these advantages come with costs. Producing matrix takes energy and raw materials. Living close to neighbors increases competition for nutrients and space. Waste products can accumulate. Viruses that infect bacteria, bacteriophages, can spread efficiently in dense communities under some conditions. The fortress is not a utopia. It is a compromise, shaped by the fact that, in many real habitats, the greatest danger is not starvation but sudden extermination.

That compromise can be felt in a simple, everyday scene: the stubborn film on the inside of a reusable water bottle. Rinse it and it looks clean. Fill it, drink from it, leave it overnight, and a faint slickness returns. The slickness is not merely residue; it is a surface community taking advantage of a stable material, intermittent nutrients, and time. If you only swish water, you remove the drifters. If you scrub the walls, you disrupt the built environment. The difference between rinsing and scrubbing is the difference between attacking plankton and dismantling a fortress.

Biofilms are often described as slimy layers, but the most important feature is not the slime; it is the way the slime allows microbes to manipulate space. Space is usually the human domain: we build walls, arrange rooms, channel water, stock shelves, control who gets in and who stays out. Biofilms do this at micrometre scales with polymers and physics. Their walls are gels. Their doors are pores. Their plumbing is channels that open and close as the matrix swells and shrinks. Their security system is not a guard but a set of gradients that make the interior a different world from the exterior.

A bacterium that joins such a structure does not become altruistic. It becomes practical. It trades some speed for a roof, some autonomy for shelter, some exposure for a buffered microclimate. In the relentless arithmetic of survival, that bargain can be extraordinarily profitable, especially in a world where the most dangerous threats arrive in sudden pulses: a brushstroke, a disinfectant, an immune flare, a dose of antibiotics. The fortress endures by being more than one thing at once: glue, sponge, filter, pantry, and rumor mill, all built from molecules small enough to ignore until they change the outcome of a life-and-death struggle on a surface.

Chapter 2

# Laying the First Stone

*To build a city, a microbe must first survive the perilous transition from free-floating nomad to anchored settler. We explore the microscopic forces that threaten to sweep bacteria away before they can gain a foothold, and the sophisticated molecular grappling hooks they deploy to secure their foundation against the elements.*

## 2.1 The Physics of Arrival

A river stone looks smooth from a distance, but up close it is a landscape: pits and ridges, patches that catch the light differently, thin films that come and go with the waterline. A microbe approaching a surface meets the same kind of landscape, only written in forces too small to feel with fingers. Whether a biofilm begins is not decided by a microbe's enthusiasm, but by a narrow zone of physics where flowing water, surface chemistry, and molecular clutter negotiate the terms of landing.

The first surprise is how rarely a surface is truly bare. Even in a freshly manufactured pipe, a catheter just removed from sterile packaging, or a ship hull scraped clean, the moment that surface touches a real fluid it starts collecting a skin. Proteins, lipids, fragments of polysaccharide, bits of dissolved organic matter, and ions arrange themselves into a thin, uneven layer. Engineers call this a *condition-*

*ing film*. The name sounds passive, like a towel left to dry, but it behaves more like a change of clothing: the surface that microbes encounter minutes later may have very little in common, chemically speaking, with the underlying steel, plastic, or enamel.

That film matters because microbes do not arrive like darts aimed at a target. In most settings they are passengers. A bacterium in a stream or bloodstream is not a strong swimmer compared with the surrounding flow; it is more like a dust mote in wind. It drifts, spins, and is swept along. The odds of it touching a surface are shaped by the same physics that governs how smoke deposits on walls or how silt settles out in a quiet bend of a river: transport by the moving fluid, and then the last micrometre problem of crossing the final gap.

That final gap is where the idea of a *boundary layer* becomes more than a textbook diagram. Fluid moving past a surface cannot slip freely at the interface; at the very surface the velocity is effectively zero. The flow speed increases with distance from the surface, creating a gradient. In a fast-flowing pipe the water in the middle may race, while the water within a few micrometres of the wall creeps. A microbe that wanders into this near-surface region enters a different world. The shear forces (the tendency of the fluid to drag it along) increase sharply with distance from the wall; closer in, the flow is gentler but also more structured, with eddies and pockets that can trap particles. The boundary layer can be a refuge from being swept away, but it is also a gatekeeper: to make contact, the microbe must pass through it.

In many natural and engineered environments, surfaces are also electrically charged. Bacterial cell envelopes often carry a net negative charge, due to phosphate groups and other acidic components in their outer layers. Many common surfaces in water, from glass to certain plastics to oxidized metals, also tend to be negatively charged. Like charges repel. If the world were only electrostatics, bacteria and surfaces would keep a polite distance, never quite touching.

But the world is not only electrostatics. Molecules attract each other at short range through van der Waals forces, the collective name for the weak attractions arising from fluctuating charges. These forces are feeble one-to-one, but across many atoms they add up, and they do not care whether the overall objects are negatively charged. So a bacterium approaching a surface experiences a push (electrostatic repulsion) and a pull (van der Waals attraction). The balance depends on distance: electrostatic effects can reach farther into the surrounding water, while van der Waals attraction becomes significant only at closer ranges. That sets up a kind of invisible topography, an energy landscape with hills and shallow valleys.

A useful way to picture it is to imagine the microbe as a tiny bead rolling toward a wall. Far away, the bead is carried by flow and random motion, barely feeling the wall. Closer in, it begins to feel repulsion, like a gentle but growing headwind. If it gets closer still, attraction begins to matter, as if the wall has become slightly sticky. Depending on the chemistry of the wall and the water, there may be a shallow resting place: a *secondary minimum*, a weak pocket of attraction at a distance of a few nanometres to tens of nanometres. A cell caught there is not firmly glued; it is more like hovering in a loose embrace. A change in flow, or a bump from another particle, can dislodge it. Yet that hovering can last long enough for the cell to do something biological: to reorient, to extend an appendage, to secrete a little adhesive. A foothold in physics can buy time for a foothold in biology.

The saltiness of the water can shift this balance dramatically. Dissolved ions screen electric fields. In low-ionic-strength water, the electrostatic repulsion between a negatively charged cell and a negatively charged surface reaches farther; in saltier water, that repulsion is muffled. This is one reason microbes behave differently in distilled water than in seawater, and why the same organism may stick readily in one setting and refuse to settle in another. It also helps explain a mundane observation familiar to anyone who has watched scale form in kettles or biofouling in plumbing: mineral-rich, ion-rich waters often encourage deposition of all kinds, living and non-living.

The conditioning film complicates the story further, because it changes not just the charge but the *surface energy*: how the surface interacts with water and with organic molecules. A clean piece of glass is famously hydrophilic; water spreads on it. But coat that glass with a thin layer of proteins and lipids, and suddenly it can present hydrophobic patches. Microbes themselves have surfaces with mixed character: parts that like water, parts that avoid it, sometimes decorated with hair-like polymers that extend outward. When two complex, patchy surfaces approach each other in a real fluid, the relevant question is not simply "is it charged?" but "what kinds of patches are exposed right now, and what does the water between them want to do?"

Water is not a passive medium. At nanometre scales it organizes. It forms structured layers near surfaces, and those layers can either resist being squeezed out or be displaced easily, depending on the chemistry. Hydrophobic areas can encourage water to retreat, bringing surfaces into closer contact; hydrophilic areas can hold onto water, maintaining a lubricating layer. A conditioning film can therefore act as a landing strip or as a non-stick coating, sometimes within micrometres of itself, depending on local composition. The film can also provide what microbes recognize as familiar: proteins resemble food, polysaccharides resemble home turf, and some bacterial adhesins bind specific host molecules. In the urinary tract, for example, proteins from urine can coat catheters quickly, transforming inert plastic into a biologically meaningful surface. In the ocean, dissolved organic matter and microbial exudates coat hulls and rocks, turning "bare" material into a chemically rich interface.

This interplay of fluid mechanics and surface chemistry has been appreciated for longer than one might guess, because it caused trouble in places that society could not ignore. In the nineteenth century, ships that spent months at sea returned with hulls thick with growth. The drag from barnacles and microbial slime could slow vessels dramatically, costing time and fuel (or, earlier, requiring larger crews and more supplies). Before anyone knew about bacterial matrices, sailors and shipwrights knew that a clean hull did not stay clean, and that some coatings delayed fouling better than others. The famous copper sheathing adopted by the British Royal Navy in the late eighteenth century was driven by performance: copper was toxic to many organisms and also altered the chemistry at the surface. It did not prevent all attachment, but it changed the odds. This was, in a practical sense, an early experiment in surface chemistry as a means of controlling biological settlement.

Modern biofilms pose similar problems on a smaller scale. In a hospital, the "hull" may be an intravenous line. In a water treatment plant, it may be a membrane. In dentistry, it is enamel. The first colonizers of dental plaque do not strike enamel as if it were a nail and they a hammer; they arrive through saliva, a fluid rich in proteins and glycoproteins that coats teeth within minutes. That salivary pellicle is a conditioning film with a name and a clinical consequence. Early plaque-forming bacteria attach not directly to enamel but to that pellicle, often by binding specific molecules within it. The physics is still there: the boundary layer of saliva, the shear from tongue and chewing, the near-surface forces. But the chemistry has been rewritten by biology before the microbes even show up.

The boundary layer also explains why some places become microbial cities and others remain relatively bare. Consider a fast-flowing stream over rock. The tops of protruding stones are exposed to strong shear; anything weakly attached is likely to be swept away. In crevices and sheltered downstream sides, the flow slows, the boundary layer thickens, and particles spend longer near the surface. Those sheltered microhabitats are the streambed's prime real estate for biofilm formation. The same principle holds in pipes, where biofilms often begin at joints, roughness, and dead zones; or in the lungs, where thick mucus and uneven airflow create pockets of slower movement. "Where a biofilm can begin" is often "where the fluid gives a cell time to negotiate the last micrometre."

Time is the overlooked currency here. A cell approaching a surface is not making a conscious decision, but it does have processes that take seconds to minutes: changes in gene expression, secretion of polymers, deployment of appendages. If the flow carries the cell past a surface too quickly, it has no opportunity to engage those processes. A shallow physical trap, a slow eddy, or a sticky conditioning film can extend contact time just enough. Engineers sometimes describe this in terms of *residence time* near the surface. Biologists see its outcome: the difference between a transient visitor and a founder.

Even the microbes' own motion, modest as it may be, interacts with the boundary layer. Many bacteria swim by rotating flagella, producing thrust that can move them through still water at tens of micrometres per second. In a rapid flow, that speed is negligible compared with the bulk movement, but near a surface where the flow slows, swimming can matter. A swimming bacterium may be pushed into a surface by hydrodynamic interactions, not unlike how a spinning ball curves in air. Close to a wall, the flow around a swimming cell becomes asymmetric, and that asymmetry can bias the cell's trajectory toward the wall, increasing the chance of contact. The same physics can also cause cells to linger near surfaces longer than expected, circulating in near-wall orbits before being carried away.

Then there is randomness. At microbial scales, Brownian motion jostles particles. For very small bacteria and for spores, this thermal jitter can be a significant component of arrival, nudging them across the last gap when flow alone might keep them slightly separated from the surface. In some settings, diffusion is the main route by which cells reach surfaces at all. In stagnant water, without stirring, a microbe may still eventually touch a surface simply because it is constantly being buffeted by molecules. This slow, patient arrival is one reason biofilms can form even in apparently quiet, clean containers left on a bench. The boundary layer is still present, but now it is not sculpted by fast flow; it is a diffusion-dominated region where the most important gradient is concentration, not velocity.

If this sounds like a delicate balance, it is. A small change in pH can alter surface charge by changing which chemical groups are protonated. A small change in ionic strength can shrink the electrostatic barrier. A trace amount of protein can transform a hydrophilic surface into a patchwork of hydrophobic islands. A bit of roughness can create sheltered pockets. No single variable explains attachment across all systems because attachment is not one event. It is a sequence of near misses and brief contacts, with probabilities that multiply.

One concrete example comes from wastewater and drinking water infrastructure, where operators sometimes observe that biofilms take off after subtle changes in water chemistry. A shift in disinfectant strategy, a seasonal change in organic carbon, or a change in water source can alter the conditioning film and the ionic environment. The early colonizers may not be the pathogens of concern, but once a foundation forms, it can create microenvironments that protect a wider community. In these systems the first settlers are not necessarily the strongest; they are the luckiest under the day's physics.

Surface chemistry can also be deliberately engineered to change the odds. Hydrophilic coatings that hold water tightly can reduce protein adsorption and make it harder for cells to come into close contact. Zwitterionic coatings, carrying both positive and negative charges, can create strong hydration layers that resist fouling. Conversely, in some beneficial applications, engineers want microbes to stick: in bioreactors that rely on biofilms to degrade pollutants, or in microbial fuel cells. There the goal becomes to create a welcoming landing strip: rough surfaces to thicken the boundary layer, chemistries that encourage adsorption of conditioning films that microbes like, and flow regimes that bring cells near without stripping them away.

All of this talk of surfaces risks implying that the surface is a single plane. In reality, at microbial scales, most surfaces are mountainous. A scratch in plastic becomes a canyon. A grain boundary in metal becomes a ridge. Even on a polished surface, nanoscale roughness creates local zones where forces and flow differ. A bacterium does not need an entire countertop; it needs one niche that fits its size. Roughness also affects the conditioning film, concentrating molecules in valleys and leaving peaks relatively exposed. This is why two samples of "the same material" can behave differently: one manufactured batch may have different microtopography, different additives that bloom to the surface, or different patterns of adsorption.

A particularly vivid case is the inside of medical devices. Consider a central venous catheter. It is designed to be smooth and inert, yet it sits in blood, a fluid crowded with proteins that are eager to adsorb. Within minutes, proteins coat the polymer. Platelets and other components may adhere and form micro-deposits. Flow within the catheter is not uniform; there are zones of slower movement near the wall, and intermittent pulses from infusion. Bacteria that enter the bloodstream transiently, perhaps from skin or a dental procedure, may pass by in huge numbers without incident. But if a few are swept into that near-wall region and find a spot where the conditioning film and ionic environment reduce repulsion, they can linger long enough to make meaningful contact. The catastrophic part is not the first contact; it is what that contact enables. A small founding population can become a persistent source of infection precisely because it is sheltered by the physics of the device and the chemistry of its coated surface.

The physics of arrival also explains an unsettling fact: killing free-floating bacteria does not automatically prevent biofilms. Disinfectants are often applied with the logic of sterilizing the fluid. But if the surface rapidly accumulates a conditioning film, and if there are protected boundary-layer niches, then the relevant battlefield is not the well-mixed bulk but the near-surface zone. Some disinfectants are consumed by reacting with organic matter in the conditioning film, weakening their effect right where it matters. Flow can replenish disinfectant in bulk while still leaving a thin zone near the surface where concentration is lower or reaction is faster. In other words, the geometry and chemistry conspire to create a small sanctuary, micrometres thick, where a founder cell can survive long enough to become hard to remove.

A final twist is that the conditioning film is not always an accidental gift to microbes. Microbes themselves can manufacture it. Even before a stable biofilm forms, free-living cells shed polymers, release DNA, and excrete metabolites. In a crowded environment, those secretions accumulate on surfaces, altering them in ways that make subsequent attachment easier for themselves or for other species. A surface becomes progressively more biological simply by being in a biological fluid. The very act of living in the vicinity rewrites the chemistry of the landing site.

So arrival is less like a plane touching down on a runway and more like a leaf settling onto a wet sidewalk in a gusty wind. The leaf might stick if there is a thin film of water and oil; it might slide if the surface is clean; it might be lifted again by a passing draft. A bacterium is smaller, the forces are different, but the logic rhymes: flow brings it close, near-surface physics decides whether it can linger, and chemistry decides whether the last nanometres are welcoming or hostile. Where those conditions align, a surface stops being merely a boundary and becomes a place where life can take hold.

## 2.2 Locking the Grapple

A bacterium that has drifted close enough to a surface to feel its pull is not yet a resident. In that first moment of contact, it is more like someone stepping onto an icy curb: balance is tentative, and a small shove from the passing flow can send it back into the stream. The earliest attachment is often *reversible*, a word that sounds calm but is really an admission of fragility. The cell is held by weak, short-range forces and by whatever molecular clutter happens to be on the surface. It can rock, roll, and slide. It can leave without tearing itself apart. Many do.

Yet biofilms exist precisely because some cells cross a threshold. They stop loitering and start fastening. That shift is biological. It depends on specialized structures—microscopic grappling gear that bacteria deploy at the right moment—and on a rapid internal decision: whether to keep investing in being a swimmer, or to become a settler.

The basic problem is mechanical. A microbe in flowing liquid experiences shear: the flow is faster a little farther from the surface than it is right at the surface, so the cell is tugged and twisted. Even if the bulk flow is gentle, the cell feels flicks and eddies and collisions with other particles. If it is attached only by weak physical interactions, those disturbances act like repeated taps on a poorly hung picture frame. Eventually the hook slips.

Bacteria solve this by turning contact into grip. One strategy is to extend thin filaments from the cell surface that can grab, retract, and grab again. Another is to present sticky proteins that bind to particular molecules on the surface. These strategies are often combined, layered like a climber using both hands and a rope.

The filaments have many names, which can make them sound more exotic than they are. *Fimbriae* and *pili* (singular: pilus) are both hair-like appendages, typically made of protein subunits assembled into long polymers. Some are stiff, some are flexible; some are primarily adhesive, some are primarily involved in movement, some do both. The key point is simple: they extend the reach of the cell, allowing it to probe and latch without committing the whole body.

It is tempting to picture a bacterium as a smooth capsule with a few hairs. Under the right microscope it looks more like a burr seed, bristling with tools. When those tools meet a surface, they do not merely stick; they can *act*. Certain pili can retract, pulling the cell toward the surface with surprising force. This is not a metaphorical tug. It is a measurable mechanical pull, powered by molecular motors inside the cell that consume ATP, the same cellular fuel that powers muscle.

Type IV pili are the best-known examples of this retractile machinery, studied in organisms such as *Pseudomonas aeruginosa*, an opportunistic pathogen infamous in hospitals and in the lungs of people with cystic fibrosis. A type IV pilus can extend outward, attach at its tip, and then retract like a winch cable being reeled in. The retraction is driven by a motor protein called PilT (among others), and the forces generated can reach on the order of $10^2$ picoNewtons per pilus. A picoNewton is an inconvenient unit for human intuition, but at the scale of a bacterium it is immense: enough to yank the cell body into tighter contact and to resist being peeled away by flow.

This retract-and-pull behaviour does two things at once. First, it converts a weak, reversible touch into a tighter mechanical embrace. Second, it changes what the cell *senses*. A surface is not just a chemical signal; it is a physical constraint. When a pilus tip sticks to a surface and the motor tries to reel it in, the pilus experiences tension. That tension is information. In *P. aeruginosa*, work in the early 2020s tied pilus retraction to a rise in an intracellular messenger molecule (cAMP), which then helps switch on biofilm-related behaviours. The details are still being worked out across different species, but the broader idea is compelling: a bacterium can detect that it is attached because its own machinery strains against something immovable.

There is an old and oddly charming historical thread behind this modern mechanosensing story. Long before anyone talked about pili, microbiologists noticed that many bacteria behave differently near surfaces. In the mid-twentieth century, as microscopy improved and researchers began filming microbes in liquid, it became clear that swimming bacteria often accumulate near walls and that some can move along surfaces in characteristic ways. Only later did the genetic and molecular tools arrive to explain the machinery behind those behaviours. What had looked like mere preference or chance turned out to be a mechanical dialogue: cells pushing, pulling, and testing the world with appendages.

A second strategy for locking on involves adhesins: surface proteins that bind to specific targets like receptors, often with remarkable strength. For a bacterium, this is a way to stop relying on the fickle balance of weak forces and instead use something closer to a molecular handshake. Many adhesins recognize host molecules. That makes sense in infection, where the most valuable surfaces are living tissues. But the same logic applies on medical plastics once they are coated with host proteins. A catheter may begin as inert polymer, but in blood or urine it quickly becomes a protein-decorated landscape that adhesins can recognize.

Here the story takes an unexpected turn, because under flow, "strong binding" does not always mean "binding that fails when pulled." Some adhesin systems behave like a peculiar kind of knot: they tighten under tension. These are called *catch bonds*, an idea that initially sounds like it must be wrong. In everyday experience, pulling on a bond usually makes it break sooner. Catch bonds do the opposite over a certain range of forces: pulling increases the lifetime of the bond.

The classic example is FimH, an adhesin found at the tips of type 1 fimbriae in *Escherichia coli*. FimH can bind to mannose sugars on host tissues. Under low shear, the binding can be relatively weak, allowing cells to roll or detach. Under higher shear, the binding becomes stronger, enabling firm adhesion even in flowing environments such as the urinary tract. Another well-studied case involves *Staphylococcus aureus* and its adhesin ClfA binding to fibrinogen, a blood protein. In the bloodstream, where forces are not gentle, a bond that strengthens under pulling is a brilliant adaptation.

Catch bonds make emotional sense once the mechanical problem is clear. If a bacterium needs to hold on in a fast-flowing environment, it is precisely the moments of stronger pull—the moments when detachment would be most likely—when it most needs reinforcement. A catch bond turns danger into opportunity. Instead of being peeled away, the cell is pressed into commitment.

The transition from reversible to irreversible attachment is therefore not a single switch but a ratchet. Weak initial contact allows a cell to sample the surface. If conditions are hostile, it can leave. If conditions are promising, appendages and adhesins begin to engage more strongly. Retraction pulls the cell closer, increasing contact area and decreasing the distance across which repulsive forces act.

Sticky proteins bind. More attachment points form. The odds of complete detachment drop quickly, because detachment would now require multiple bonds to fail at once.

This ratchet also explains why early biofilm formation is often heterogeneous. Even genetically identical cells in the same flow can behave differently. One cell may touch down in a sheltered pocket, long enough to deploy pili and begin retraction. Another may make contact on a more exposed patch and be swept away before it can engage. A third may carry slightly different surface decorations at that moment—different expression levels of appendages, different outer-layer molecules—and so the balance tips in or out of attachment. The founding of a biofilm can begin with a few cells that are not necessarily special in any grand sense, but happen to cross the threshold first.

There is another, quieter part of the biological shift: the cell starts shutting down its swimming machinery. Swimming is expensive. Flagella are complex motors and propellers that require constant energy to run and constant maintenance to build. A cell firmly attached to a surface gains little from continuing to swim, and the spinning of flagella can even interfere with stable attachment by producing torque. In many species, surface contact triggers regulatory changes that slow or stop swimming and promote adhesive behaviours. The cell becomes less of a nomad and more of a builder.

That change is not merely economical; it alters the cell's relationship to the local environment. A swimmer experiences the world as rapidly changing—a new chemical landscape every second as it moves through gradients. An attached cell experiences the world as persistent—the same patch of surface chemistry, the same local flow, the same neighbours. That persistence makes long-term investments sensible: producing adhesion structures, secreting polymers, coordinating with nearby cells. The first firm bonds are therefore also the first steps toward a different kind of life.

A concrete example of this shift can be found in chronic lung infections in cystic fibrosis. The thickened mucus and altered chemistry of the airway create surfaces where bacteria can persist. *Pseudomonas aeruginosa* is notorious for establishing biofilms in this setting. Early in colonization, individual cells must hold on despite coughing, airflow, and immune activity. Type IV pili contribute both to movement along surfaces (twitching motility) and to firm attachment through retraction. Once attached, cells can transition into biofilm growth, becoming far harder to clear. The tragedy, clinically, is that the earliest mechanical decisions made by a few cells can echo for months or years in the form of persistent infection.

The same mechanical logic plays out in non-medical settings that touch everyday life. In food processing equipment, bacteria must attach to stainless steel surfaces that are regularly washed and subjected to flow. On ship hulls and submerged infrastructure, the earliest microbial films must resist waves and currents. In water pipes, early colonizers must withstand bursts of flow. In each case, reversible contact is cheap; irreversible attachment is an investment that pays off only if the cell can keep its grip long enough to begin modifying the local environment.

A useful way to see this is to imagine two different failure modes. A reversibly attached cell fails by slipping away, like a suction cup that never seals. An irreversibly attached cell fails by being torn, peeled, or broken away, which requires more force and often leaves remnants behind. The biological structures that promote irreversible attachment are, in essence, solutions to peel forces. Multiple pili distribute load. Adhesins bind in ways that resist shear. Retraction pulls the cell into close contact, reducing leverage that would otherwise allow the flow to pry it off.

The physical picture becomes clearer when the forces are drawn. Thick arrows for flow-induced drag pulling the cell downstream; thinner arrows for adhesive interactions pulling it toward the surface; a retraction arrow showing a pilus shortening and tightening. The transition from reversible to irreversible attachment is the moment when the arrows change relative size: the pull toward the surface becomes dominant, not because the flow has stopped, but because the cell has begun to fight back with molecular machinery.

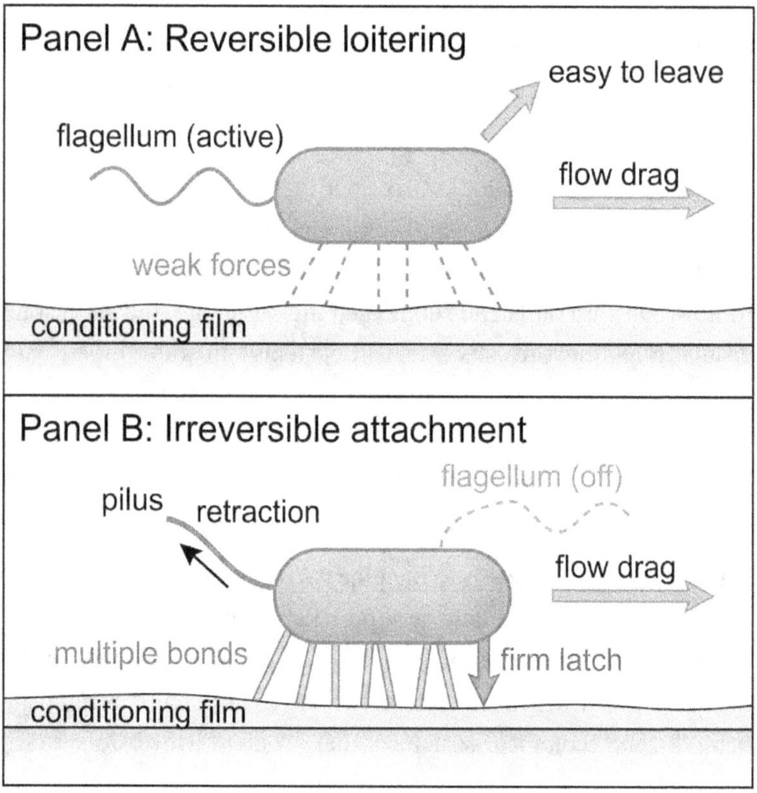


One reason this biological lock-in is so effective is that it recruits feedback. A cell that pulls itself closer increases the chance that more adhesins will find their targets. More bonds reduce motion, making it easier for still more bonds to form. The cell is also not alone for long. Once one bacterium becomes a stable anchor, it changes the local flow ever so slightly and provides a surface that may be easier for other cells to stick to than the original material. Many bacteria adhere readily to other bacteria, especially once sticky polymers and surface proteins begin to accumulate. The first firmly attached cell is not merely a resident; it is a scaffold.

There is a moral, in the literal sense of a lesson about what matters, buried in this microscopic mechanics. People often imagine microbes as chemically clever but physically simple. The truth is more unsettling: they are also mechanically ingenious. They build retractable cables, force-sensitive bonds, and surface-sensing systems that let them convert an accidental brush with a wall into a stable, defended address.

A faucet handle, a contact lens case, the inner wall of a ventilator tube, the enamel of a tooth, a river rock slick with algae—these look like solid objects to us. To a bacterium they are dynamic interfaces, places where forces rise and fall and where a tiny grappling system can decide whether the cell becomes a passing speck or the first tenant of a crowded, persistent community. The decisive moment is not the first touch, but the instant the cell begins to pull back.

## 2.3 Manufacturing the Mortar

A lone cell that has managed to fasten itself to a surface is still terribly exposed. Flow can tug at it, disinfectants can wash over it, immune cells can find it, and even a small change in chemistry can turn yesterday's welcoming patch into today's hostile ground. The most important move a founder cell can make is not to recruit neighbours, but to change the material rules of the place it has claimed. It does that by building a matrix: a secreted, sticky substance that turns a single attachment point into a sheltered address.

Microbiologists call the matrix *extracellular polymeric substances*, abbreviated EPS. The name is deliberately broad. "Extracellular" means outside the cell; "polymeric" means made of long-chain molecules; "substances" is a polite way of admitting that this is a messy mixture rather than a tidy chemical. EPS is not one thing but a composite: sugars knitted into long polysaccharides, proteins of many kinds, lipids, and a component that still sounds surprising to many non-specialists—DNA outside cells, called extracellular DNA or eDNA.

The most useful first image is mortar. Not because a biofilm is a wall of bricks, but because mortar is what turns separate stones into something that behaves like a single object. If you kick an unmortared pile of stones, the stones scatter. Kick a mortared wall, and the wall pushes back. EPS is the difference between a set of individual microbes stuck near each other and a community that can endure.

EPS also changes what it means to be attacked. A free-swimming bacterium experiences an antibiotic molecule as an immediate personal threat: the drug diffuses to the cell, binds its targets, and interferes with growth or kills. In a biofilm, the same antibiotic arrives at the outer edge first, meets a dense matrix of polymers, and begins a longer negotiation. Some drugs are slowed simply because the path is tortuous. Some are trapped or inactivated by chemical interactions with matrix components. Others penetrate but meet cells in altered physiological states. The matrix is not merely a barrier; it is a staging ground where chemistry and biology reshape each other.

That reshaping begins early, sometimes within minutes of irreversible attachment. A surface-attached cell secretes polymers that remain nearby because flow near the surface is slow and because the polymers themselves are sticky. These secretions form a kind of glue halo around the cell. The halo catches other cells like lint in a cobweb. It also catches dissolved molecules and particles from the surrounding fluid. The first EPS is therefore both construction material and net.

Polysaccharides, the sugar-based part of EPS, are often described as "sticky," but their stickiness comes from specific physical features. A long polymer chain in water behaves like a flexible thread that can tangle with others. Many polysaccharides also bind water strongly, forming hydrated gels. A hydrated gel is not simply wet; it is a structured network in which water is held in place by the polymer scaffold. This matters because a gel can be soft yet resilient. Under shear, it can deform instead of tearing. When shear relaxes, it can recover. These are good properties for a community living in flow.

Proteins in the matrix do many jobs. Some are enzymes that remodel the matrix itself—cutting and rejoining polymers, adjusting stiffness, freeing nutrients. Some are adhesins that help cells bind to each other or to the surface. Some are structural fibers that add tensile strength, more like rebar than glue. In some bacteria, proteins assemble into amyloid-like fibers that are remarkably tough; in others, proteins act as crosslinkers between polysaccharides or between cells and the matrix.

Extracellular DNA deserves special attention, because it upends a familiar idea. DNA is often introduced as the intimate genetic library inside cells. Seeing it outside cells feels like seeing books scattered in the street. Yet eDNA is common in biofilms, and it is not just debris. DNA is a long, negatively charged polymer. Long and charged means it can behave like a scaffold. It can entangle with other polymers. It can bind positively charged molecules. In some systems eDNA is important very early, contributing to initial cohesion, and experiments where DNase (an enzyme that cuts DNA) is added can disrupt young biofilms. That is a dramatic demonstration: a biofilm that looks like a continuous smear can be weakened by an enzyme that simply snips a polymer backbone.

Where does the eDNA come from? Sometimes from cell death, which is not always accidental. In many microbial communities, a fraction of cells lyse and release contents that benefit the group. Sometimes DNA is actively exported or released in vesicles. Sometimes it comes from the environment. The result is the same: DNA becomes a building material. It is one of the clearest examples of how biofilms blur the boundaries between individuals. A molecule that once served a private purpose becomes public infrastructure.

A concrete, human-scale example of EPS as mortar is dental plaque. Teeth in a mouth are repeatedly washed, scraped, and chemically challenged. Yet plaque persists. Part of the reason is mechanical: the matrix holds water, and that hydrated gel adheres to the pellicle-coated enamel and to itself. Part is ecological: plaque is a community, with microbes that feed on the sugars we eat and microbes that feed on the metabolites produced by others. But the stubbornness we experience when flossing is, in a direct sense, a material property of EPS. It is hard to remove because it is not merely a sprinkling of cells; it is a polymeric fabric binding cells together and binding them to the surface.

The same material logic has consequences far from the mirror. In chronic infections, biofilm matrix can shield bacteria from immune attack. Immune cells such as neutrophils are large compared to bacteria. They cannot easily penetrate tight polymer networks. They release reactive molecules and enzymes, but those can be quenched or diverted in the matrix. Even when immune cells succeed in killing bacteria at the biofilm edge, deeper cells can persist. Clinically, this is part of why infections involving implanted devices can be so difficult to eradicate without removing the device itself: the matrix provides the microbes with a protected microhabitat on a surface that the body cannot simply shed.

The historical anecdote here is less a single dramatic moment and more a slow recognition. For much of microbiology's early history, bacteria were studied in well-mixed liquids. They grew as free-floating cells in flasks and test tubes because those were easy to handle, easy to measure, and easy to standardize. The idea that bacteria commonly live in attached, matrix-embedded communities gained momentum in the late twentieth century as microscopy and clinical observations converged. When researchers began to look closely at surfaces in natural waters, industrial systems, and infected tissues, they saw that bacteria were not primarily solitary swimmers. They were builders. The EPS matrix, once treated as an inconvenient slime, became the central fact.

Calling EPS "slime" is not wrong, but it misses an important point: slime is not a substance with one formula. Biofilm matrix composition varies across species and even across strains within a species. A particularly well-studied case is *Pseudomonas aeruginosa*, which can produce multiple matrix polysaccharides, including polymers often referred to as Pel and Psl. Work from Roberto Kolter's group in 2004 helped establish that these are distinct carbohydrate-rich components with structural roles. Later studies revealed an elegant chemical detail: Pel is cationic, meaning it carries a net positive charge, and it can cross-link with negatively charged eDNA through ionic interactions. That is a simple physical principle—opposites attract—turned into architecture. A positively charged polymer can bind a negatively charged polymer, creating a reinforced network that is stronger than either component alone.

This kind of cross-linking is the microbial version of mixing concrete with reinforcing fibers. The result is a matrix that is harder to deform and harder to break. It also changes diffusion, slowing the movement of some molecules while allowing others through. Importantly, it can create microenvironments. Oxygen can be depleted inside a thick biofilm because outer cells consume it faster than it diffuses inward. Nutrients can form gradients. Waste products can accumulate. These gradients matter because bacteria respond to them, altering their metabolism and their susceptibility to drugs. Material structure and physiological state become intertwined.

EPS production is also a budgeting problem. Polysaccharides are built from sugar precursors that must be synthesized or imported. Making large amounts of polymer is expensive. In *P. aeruginosa*, certain pathways draw on shared precursor pools; pushing hard on one polymer can reduce another without any simple on-off switch at the gene transcription level. This is a reminder that microbes are not merely following blueprints. They are managing resources in real time, under constraints, responding to what the environment offers.

One consequence is that biofilms can differ in texture and toughness depending on conditions. A biofilm grown under high shear may invest in a matrix that resists deformation and holds cells tightly. A biofilm in a nutrient-rich but low-shear environment may spread into a more gelatinous layer. A biofilm exposed to intermittent drying may produce polymers that retain water. These are not conscious choices, but evolved responses encoded in regulatory networks that connect surface sensing, nutrient availability, and stress signals to matrix production.

The matrix is not only defense; it is communication infrastructure. Polymers can hold signaling molecules near cells, increasing local concentrations. They can bind metals, altering availability. They can trap enzymes, keeping digestive tools close to where they are useful. Even the physical proximity enforced by a matrix changes the social life of microbes. In a well-mixed liquid, two neighbouring cells may be separated a moment later. In a matrix, neighbours remain neighbours long enough for cooperation, competition, and chemical warfare to become stable strategies.

A vivid way to appreciate how quickly matrix changes a community is to think about what happens when a second cell arrives. Without EPS, two cells near each other may collide and separate, like billiard balls in slow motion. With EPS, a collision can become a capture. A polymer strand can bridge the gap, and now the two cells share a local microenvironment. If one cell releases enzymes that free nutrients from the surface or from the conditioning film, the other benefits. If one cell alters pH locally, the other experiences it. The matrix turns mere proximity into shared fate.

EPS can also make biofilms deceptively strong against physical cleaning. A flowing rinse may remove free cells but leave behind a thin polymeric film that still contains microbes. That residual film can seed regrowth. In food production or hospital settings, this is one reason why surface cleaning is not simply a question of killing microbes; it is also a question of removing matrix. Some cleaning regimens combine surfactants (to disrupt hydrophobic interactions), oxidants (to attack organic matter), and enzymes (such as proteases or DNases) to break the matrix into pieces that can be washed away. The engineering problem mirrors the biological one: the matrix is a composite, so dismantling it often requires multiple tools.

None of this implies that EPS is invincible. Biofilms can be disrupted by changes in flow, by mechanical abrasion, by chemical treatments, or by biological antagonists. Many microbes produce enzymes that degrade other microbes' matrices. Some bacteriophages carry matrix-degrading enzymes that help them penetrate biofilms. In nature, biofilms are constantly being built and eroded, like cities exposed to weather and conflict. But the existence of effective countermeasures is itself evidence of how important the matrix is. Evolution does not build elaborate tools to defeat something trivial.

At the scale of the individual cell, manufacturing mortar is a leap of faith with a practical payoff. A cell invests energy and material into polymers that, strictly speaking, are outside its body. It is building something it cannot fully control. A passing flow could still tear it away. A competitor could exploit it. Yet if the cell succeeds in turning a few micrometres of space into a gel, the local physics changes in its favor. The surface becomes stickier, the flow becomes less scouring, and the chemical world becomes more buffered.

At the scale of a human life, these micrometres matter. The difference between a clean medical implant and a chronic infection can begin with a thin film of polymer that a founder cell extrudes in the first hours of attachment. The difference between a stable water system and recurrent contamination can hinge on whether early biofilms are allowed to lay down a matrix in dead zones of piping. The difference between healthy gums and periodontal disease begins with a film so thin it is hard to see, yet tough enough to resist a toothbrush unless the bristles and chemistry are right.

What is most striking, once one gets used to thinking this way, is how physical the earliest stages of microbial community life really are. The first colonists do not simply land and multiply. They change the material properties of their surroundings, making a private adhesive into public architecture. Polymers, charges, and water become the language of permanence, and a surface that once offered only a precarious foothold becomes a foundation with its own weatherproofing, its own scaffolding, and its own capacity to hold.

## Chapter 3

# Skyline and Infrastructure

*Biofilms are not distinct organisms, but rather sophisticated cities with towering architectures and functioning infrastructure. This chapter descends into the microscopic skyline to reveal how bacteria engineer their environment, building circulatory systems and diverse neighborhoods that redefine what it means to be a single-celled organism.*

## 3.1 Towers, Mushrooms, and Waterways

A young biofilm often looks disappointingly plain. If you could shrink yourself to microbial scale and stand on the surface where the first cells have attached, you might see a thin, slick film, more like a fogged window than a city. Then, without any architect issuing blueprints, the flatness breaks. The carpet buckles upward. Clumps thicken into ridges. Bulges become domes. With time and the right conditions, the domes sharpen into towers and the towers develop caps, producing the classic "mushroom" silhouettes that made early biofilm microscopists stop and stare.

The remarkable part is not just that bacteria can build in three dimensions, but that the shape is not cosmetic. The skyline solves a problem that every crowded community faces: how to feed the interior without poisoning it. A dense microbial population is a chemical engine. Cells consume oxygen and nutrients; they exhale acids, gases, and other wastes; they leak enzymes and polymers; they respond to all of this by changing their behavior. In a well-mixed liquid, currents do the housekeeping. In a biofilm, the cells have chosen immobility (or at least, the appearance of it). If nothing moved through the community, the biofilm would choke on its own success. The fact that biofilms can grow thick and persistent tells you that they have, in effect, invented infrastructure.

The first clues came from a deceptively simple question: is a biofilm a uniform slab, like a layer of paint, or is it lumpy and porous? For a long time, microbiology encouraged the "smear" view of microbial life. Spread cells on a flat surface, stain, look through a microscope: you see dots and rods arranged in a two-dimensional plane. Biofilms refused to behave like that. When researchers began imaging intact biofilms in hydrated form, especially using confocal laser scanning microscopy in the early 1990s, they encountered a landscape that was closer to coral than to paint: dense aggregates separated by open voids, with channels that looked suspiciously like plumbing. Costerton and colleagues were among those who championed the idea that biofilms create "customized microniches," meaning that the chemistry inside can diverge sharply from the surrounding environment over distances smaller than the thickness of a human hair. That concept is hard to appreciate until you see the architecture that makes it possible.

A mature biofilm is not a single object. It is a composite of at least three interacting materials: cells, the extracellular polymeric substances (EPS) they produce, and the water that threads through everything. EPS is often described as slime, which is true in the way that calling concrete "mud" is true. EPS is a mesh of sugars, proteins,

DNA, and other polymers that can bind water, stick to surfaces, and hold neighbors close. It glues the community together, but it does not necessarily fill all space. Biofilms are famously hydrated. Much of what looks like "empty" space in images is actually water: the medium through which nutrients arrive and wastes depart.

The mushroom form brings these ingredients into a recognizable pattern. Imagine a stalk rising from the surface: a dense column where cells and EPS are tightly packed. Above it, a cap spreads laterally, often broader than the stalk beneath it. Between neighboring stalks and caps lie channels—gullies, alleys, and sometimes cavernous voids—filled with moving fluid when conditions allow. The cap-and-stalk structure is not universal, but it is common enough to have become an icon, especially in the biofilm of *Pseudomonas aeruginosa*, a bacterium that can inhabit soil and water but also causes stubborn infections in hospitals and in the lungs of people with cystic fibrosis.

That icon, however, can mislead if it suggests a static monument. Mushroom structures can be assembled by different cellular subpopulations that behave differently even when genetically identical. One vivid demonstration came from work in the early 2000s by Klausen and colleagues studying *P. aeruginosa* biofilms. They observed that the stalk and cap could arise through a division of labor: relatively non-motile cells proliferated locally to create a base, while more motile cells migrated and accumulated higher up, contributing to the cap. The motility involved is not the familiar swimming through liquid but a surface-associated movement driven by type IV pili, tiny retractable filaments that pull the cell forward in a jerky crawl known as twitching motility. In other words, at least some of the skyline is built by builders who can climb.

The historical anecdote here is not just about an organism but about a change in what microbiologists considered visible. Before confocal microscopy became widely used, a thick biofilm was often studied by scraping it off, homogenizing it, or slicing it physically. Those approaches reveal what is present but destroy the spatial logic of how it is arranged. Confocal imaging made it possible to watch three-dimensional structures develop without dismantling them, and it did so in a way that respected the biofilm's natural state: wet, translucent, and uneven. The mushroom towers were not a metaphor imported from urban fantasy; they were what the data showed.

Once you accept that the biofilm has hills and valleys, a second question becomes unavoidable: what are the valleys for? Some early interpretations treated the voids as accidents, gaps left behind by uneven growth. That view became harder to defend when researchers started measuring transport. Oxygen microelectrodes and other microsensors, paired with imaging, revealed that oxygen penetration depends not only on how thick the biofilm is, but on how it is shaped. A flat slab forces molecules to diffuse straight down, and diffusion is slow over distance. But a landscape carved with channels increases the surface area in contact with flowing liquid and creates shorter pathways from water to cells. In some systems studied by de Beer and Stoodley in the mid-1990s, internal voids supplied a substantial fraction of the oxygen that the biofilm consumed. The gaps were not mere absence; they were part of a delivery network.

The most striking evidence for infrastructure came when flow itself was observed inside biofilms. In everyday life, we tend to equate "transport" with a current that carries material from one place to another. In a small, viscous world, diffusion can dominate: molecules wander from high concentration to low, slowly but inexorably. Biofilms exploit both. Fluid moving in channels can bring fresh nutrients close to the biomass (convection), while diffusion carries those nutrients from the channel into the dense matrix and ultimately into cells. At the same time, waste products diffuse out to the channels, where flow can whisk them away. This division of labor between convection and diffusion is the microbial version of having highways and sidewalks.

It is tempting to call the channel network a circulatory system, but it is a peculiar one. There is no heart, no valves, no deliberate pumping. The driving forces are external: the flow of liquid over the surface, pressure differences in a porous environment, or disturbances that create local currents. Yet the effect can resemble circulation. The biofilm's shape can direct flow, create protected zones of low shear, and maintain open pathways that would otherwise fill in. Biofilms are not passive obstacles in a stream; they are material objects that reshape the stream, and in reshaping it, they reshape their own future.

A concrete example makes this less abstract. Consider a catheter or a stent in a medical setting. Fluids move through these devices under pressure, carrying nutrients and oxygen. A biofilm that forms on the inner surface is exposed to shear forces from the flow. At first glance, this seems like an advantage for the patient: perhaps the flow will simply wash the microbes away. Instead, biofilms often respond by becoming more structured and more tenacious. The EPS matrix acts like a glue and a shock absorber, distributing forces across the community. As the biofilm thickens, it can create sheltered regions where shear is low—microbial cul-de-sacs—which allow further accumulation. In some flow environments, biofilms form filamentous "streamers" that extend into the channel like waving kelp. Drescher and colleagues (2013) showed how such streamers can lead to sudden clogging, not by slow uniform narrowing but by a kind of catastrophic capture: matrix material and cells shed into the flow, forming a net that snags more material, accelerating blockage. Infrastructure, in this case, has a dark side. The same capacity to channel flow and trap resources can transform a medical device into a blocked pipe.

Even in less dramatic settings, the interplay between architecture and flow is a continual negotiation. A biofilm must remain attached while allowing enough exchange to avoid suffocation. Towers help with attachment and defense: a dense core anchored to the surface is hard to dislodge, and the EPS-rich interior can slow the penetration of toxins. Channels help with survival: they prevent the interior from becoming chemically isolated. But these goals can conflict. A too-dense matrix can hinder transport; a too-open structure can become vulnerable. The skyline is a compromise shaped by local conditions: nutrient levels, flow rate, surface chemistry, and the genetic toolkit of the microbes involved.

Why do the towers often look like mushrooms rather than, say, smooth cones? One reason is that growth tends to amplify small advantages. If a slight bump receives a bit more nutrient-rich flow or has a bit more access to oxygen, it may grow faster, becoming a larger bump that captures even more resources. The cap shape can emerge because once a stalk rises into better-fed space, lateral expansion becomes easier: the outer surface is now closer to resources, and new cells can proliferate outward while remaining connected. Meanwhile, cells in the interior may be starved and slow, contributing to the stability of the structure by not changing much. The mushroom is a record of unequal opportunity written in biomass.

Another reason is that not all cells behave the same way, even when they are clones. If a subset produces more EPS, it may create a sticky scaffold. If another subset is more motile, it can explore and colonize the top. If yet another subset is primed for stress tolerance, it may preferentially occupy the interior where diffusion is slow and toxins linger. This physiological heterogeneity is often framed as a response to chemical gradients, but it is also a driver of shape. Builders and residents are not identical roles. The skyline emerges from their interactions.

At this point, a skeptic might ask whether channels are always real or whether they are artifacts of imaging: spaces created when a biofilm shrinks during preparation or when staining fails to label certain regions. That skepticism was healthy, and it forced the field to triangulate using multiple methods. Scanning electron microscopy can distort hydrated structures; physical sectioning can compress them; fluorescent stains can bind unevenly. The compelling case for genuine channels came from the convergence of techniques and from measurements of function. When different preparation methods reveal similar voids, and when microsensors detect oxygen profiles consistent with enhanced transport near those voids, and when particle-tracking methods detect flow within them, the simplest explanation is that the voids exist as part of the living system.

The idea of "waterways" inside a biofilm also corrects a common misconception: that EPS forms an impenetrable shell. EPS can be protective, but protection does not require total sealing. A raincoat keeps you dry not because it has no pores at all, but because its structure and chemistry control how water passes through. Similarly, biofilms can control permeability. Some regions may be gel-like and slow to exchange; others may be open conduits. The result is not uniform resistance but a patchwork of barriers and corridors.

That patchwork has consequences that reach beyond simple feeding. A channel network can act as a distribution system for signaling molecules, the chemical words of microbial conversation. If a quorum-sensing signal diffuses through a dense matrix, its reach is limited and its timing slow. If channels carry fluid, signals can spread rapidly along certain routes, creating neighborhoods that coordinate internally while remaining out of sync with other neighborhoods just micrometers away. The same applies to toxins, enzymes, and even bacteriophages. In a biofilm, who hears what, and when, can depend on the architecture as much as on the genes.

Architecture also changes the arithmetic of exposure. Many antimicrobial treatments are designed with the implicit assumption that the target is a uniform population in a well-mixed environment. A mushroom biofilm breaks that assumption. The cap surface is exposed to the outside world and may be readily attacked, but the stalk core can be shielded, and the channel geometry can create zones where a drug is diluted or rapidly washed away. Conversely, some pockets may trap a compound, prolonging exposure. These uneven patterns help explain why biofilms are tolerant to antimicrobials even when their cells are not genetically resistant. Tolerance here is a property of the city, not just of its citizens.

A useful way to picture the skyline is to think not of a single tower but of an entire district whose shapes interact. Towers alter the local flow field. A protrusion can create eddies, stagnant zones, and preferential currents that feed certain channels while starving others. Over time, this can lead to feedback: regions that receive more flow remain open; regions that receive less may fill in with biomass or collapse. The biofilm becomes a living landscape shaped by fluid dynamics. Nadell and colleagues (2017) demonstrated how matrix production and flow environment can shape competition among strains: in some regimes, matrix-producing cells can obstruct flow and create low-shear refuges that allow other variants to persist. Even when the inhabitants compete, the built environment can enforce coexistence.

All of this makes the skyline feel strangely familiar. Cities, too, grow through feedback between infrastructure and behavior. Roads attract traffic; traffic demands roads. Dense neighborhoods need water and waste removal; water and waste removal allow further density. Biofilms compress that logic into millimeters and minutes. They do not plan; they respond. But response, repeated and accumulated, can produce structures that look purposeful.

The biofilm skyline also invites humility about scale. A water channel a few micrometers wide can sustain thousands of cells along its edge, and a tower only tens of micrometers tall can contain steep gradients in nutrients, oxygen, and pH. The distances are so small that diffusion should be enough, and yet biofilms still invest in channels. That tells you something about how fiercely life competes for resources. When consumption is intense, even micrometers matter. When waste is toxic, even a short delay in removal can be deadly. The architecture is a solution to crowding that emerges from the physics of transport and the biology of growth.

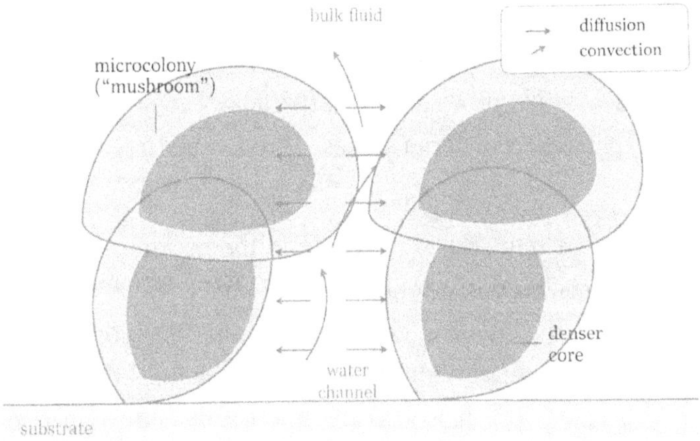


In this diagram, the arrows exaggerate to make a point that is easy to miss: most of the biofilm is not "in" the water, but it depends on water moving nearby. Convection is the courier that brings fresh supplies along the channel; diffusion is the quiet seep that delivers those supplies into the dense interior where cells actually live. The channel is not a river running through open air; it is a narrow, fluid-filled space constrained by soft walls that can swell, contract, and shed material. Its shape can change as cells grow and as EPS accumulates, which means the plumbing is always under renovation.

Renovation can be driven by something as mundane as a change in food. In laboratory systems, small shifts in nutrient concentration and inoculation density can alter whether mushroom-like towers form at all. Under richer conditions, growth may be so fast and widespread that the biofilm becomes more uniform, with fewer pronounced towers. Under leaner or more spatially patchy conditions, localized foci can gain an advantage, building upward and outward. Even in nature, this sensitivity matters. A biofilm on a river rock experiences fluctuating flow, temperature, and nutrient pulses. A biofilm in a drinking-water pipe experiences long periods of low nutrients punctuated by brief surges. The skyline is a record of those histories.

There is also a human lesson hiding in the microbial engineering. When biofilms become problems in hospitals, industry, or public infrastructure, the intuitive response is to attack the biomass directly: scrape, disinfect, blast with antibiotics. Sometimes that works; often it does not. The reason is that the architecture is part of the defense. A mature biofilm is not merely a pile of cells; it is a built environment that manages transport, creates shelters, and distributes stress unevenly. Any attempt to remove it must contend not only with the microbes but with their roads and water channels, their concrete and rebar. The skyline is not decoration. It is how the city keeps breathing and keeps being fed, even when the outside world would prefer it starved.

## 3.2 The Geography of Breath

Breathing feels like the most straightforward thing in the world: air goes in, carbon dioxide comes out, repeat. Yet even for humans, breath is geography. A few thousand meters of altitude turns a brisk walk into a slow trudge. A smoky room changes a conversation into a cough. In a biofilm, those changes happen over distances so small that a dust mote would seem enormous. A bacterium can move the length of its own body and cross an invisible border from plenty to deprivation, from oxygen-rich to oxygen-starved, from energetic growth to survival mode. The skyline matters here, not because it is pretty, but because height and depth create gradients—steep, crowded slopes in chemistry.

A gradient is simply a change across space: more of something here, less of it there. In a well-stirred glass of water, gradients are quickly erased by mixing. A biofilm is what happens when mixing is replaced by structure. Dense clusters of cells consume oxygen and nutrients faster than those molecules can be replaced by diffusion alone. The sticky matrix slows movement, not by acting like a sealed wall, but by turning a free-flowing environment into a labyrinth. Channels can carry fresh water close to the surface of a microcolony, but once molecules enter the biomass they still have to wander through polymer and crowded cells. The result is that a biofilm can contain multiple chemical climates at once.

Oxygen is the best-known example because it is both precious and easy to measure. Many bacteria use oxygen as the final electron acceptor in respiration, extracting a large payoff of energy from each unit of food. Remove oxygen and the same organism may switch to alternative electron acceptors (such as nitrate), or to fermentation, which yields far less energy but can keep the cell alive. If none of those options is viable, cells may slow almost to a halt, entering dormant states that are metabolically quiet but stubbornly persistent. The striking point is that these choices are often not genetic. They are local. Two cells with the same DNA, born minutes apart, can find themselves living different lives because one is perched near a breathable interface and the other is walled into the foundation.

Imagine standing at the outer surface of a mature mushroom-shaped microcolony. Water flows past, carrying dissolved oxygen. At that boundary, oxygen concentration is close to whatever is in the bulk fluid. Now step a few micrometers inward—a distance comparable to a bacterium's width. The oxygen has to diffuse through the matrix and around other cells, and every neighboring cell is consuming it. If consumption is rapid, oxygen can plunge from abundant to nearly zero over ten or twenty micrometers. That is not a slow fade but a cliff. In the 1990s, work by de Beer and Stoodley combining microsensors with structural imaging helped make this point unavoidable: oxygen distributions are tightly coupled to architecture. A biofilm is not a uniform blanket where oxygen falls smoothly with depth; it is a topographic map where local peaks and valleys, pores and dense cores, determine who gets to breathe.

That coupling has a surprising consequence. "Depth" is not destiny. It is tempting to think that cells near the top always have oxygen and cells near the bottom never do. In reality, a deep channel can deliver oxygen laterally to a sheltered interior, while a surface pocket in a dense cap can become oxygen-depleted because nearby cells consume faster than diffusion can supply. The gradients are three-dimensional. A cell might be close to the surface in terms of height, yet functionally buried behind a wall of hungry neighbors. Another might sit deeper but adjacent to an open void that acts like a ventilated alley. The city metaphor earns its keep here: urban air quality is not determined solely by distance from the edge of town; it depends on street canyons, wind corridors, and where traffic concentrates.

Biofilms have their own versions of those effects.

The phrase "customized microniche" captures what this does to microbial identity. A microniche is a tiny habitat with its own rules. In a forest, niches might be sunny clearings and shaded undergrowth. In a biofilm, niches are chemical: oxygen-rich skin, low-oxygen middle, anoxic base; acidic pockets beside neutral zones; regions rich in one nutrient because of a nearby channel, regions starved because they lie behind a diffusion barrier. Costerton and Lewandowski emphasized early on that biofilms do not merely occupy a surface; they reshape local conditions so strongly that the biofilm becomes its own environment. The biofilm is not just in the water. The water is, in a sense, in the biofilm.

How do genetically identical cells respond? They do what living systems have always done when confronted with gradients: they specialize. Not by adopting permanent castes, but by switching metabolic programs in response to local cues. The term "metabolism" can sound forbidding, but it is simply the set of chemical reactions a cell runs to harvest energy and build new material. Different metabolisms use different tools, and those tools are expensive to maintain. If oxygen is plentiful, a cell may invest in enzymes for aerobic respiration and grow quickly. If oxygen runs out, it may shut those down and activate other pathways. If food is scarce or stress is high, it may reduce activity, repair damage, and wait.

The smallest practical consequence is that growth becomes patchy. A biofilm that looks solid to the naked eye can contain zones of rapid cell division adjacent to zones of near-stasis. This matters for almost anything we try to do to a biofilm. Antibiotics often target active processes: building a cell wall, copying DNA, assembling proteins. A cell that is barely growing can be less vulnerable, not because it has a special resistance gene, but because it has temporarily stopped doing the things the drug attacks. That is one reason biofilm infections can be so difficult to eradicate. The drug may clear the metabolically active "surface dwellers" and leave behind a quiet foundation that can repopulate later.

A more subtle consequence is that the same chemical gradients that force specialization also encourage cooperation and conflict. If one zone runs low on oxygen, fermentation may produce organic acids or alcohols that diffuse outward. Neighboring cells in more oxygenated regions might consume those byproducts, turning waste into food. Alternatively, fermentation may acidify a pocket, stressing everyone nearby. Cells that can tolerate low pH may gain an advantage. Even without invoking social behavior, the physics of diffusion makes cells interdependent. What one cell consumes or produces shapes the neighborhood. That is why biofilms so often behave like communities rather than crowds.

There is a historical reason oxygen became the emblem of biofilm stratification: it could be measured with exquisite precision using microelectrodes small enough to penetrate the biofilm without destroying it. Those measurements were not just technical triumphs; they were conceptual turning points. They turned vague ideas about "limited penetration" into hard numbers and sharp profiles. Researchers could see, in micrometers, where oxygen collapsed. They could correlate that collapse with visible structure: a dense cluster versus an open channel. They could watch the profile change when flow rate changed, when nutrients changed, when the biofilm thickened. The geography of breath became a map rather than a metaphor.

The map is not always intuitive, and that is partly because oxygen is a moving target. Bacteria consume oxygen faster when food is abundant. If nutrients surge, oxygen may be stripped quickly near the surface, starving deeper regions even more severely. If nutrients are scarce, oxygen may penetrate deeper simply because consumption is lower. This means that the oxygen profile is not a static property of thickness alone; it is an outcome of the balance between supply and demand. The same biofilm can be "well ventilated" at one moment and "suffocating" at another.

Flow complicates things further. Water channels can carry oxygen-rich fluid into the interior, but only if they remain open and connected to the bulk. A blocked or collapsed channel turns a ventilated corridor into a dead end. Meanwhile, high shear at the surface can thin the biofilm, keeping it from growing into anoxia, while low shear can allow thick, stagnant layers to accumulate. The built environment and the external environment negotiate continuously.

Even within a single microcolony, the consequences of an oxygen gradient can feel almost theatrical. At the surface, cells can behave like exuberant entrepreneurs: growing, dividing, producing matrix, competing for space. Deeper in, cells may become conservative: the same genes, the same lineage, but a different set of priorities. In the deepest, oxygen-free pockets, cells may run on alternative chemistry or enter a persister-like state, remarkably hard to kill. The biofilm thus contains multiple tempos of life layered together.

It is worth pausing on a common misunderstanding: oxygen limitation does not mean the biofilm is dead inside. Anoxic does not equal inactive. Many microbes thrive without oxygen. The catch is that they require different electron acceptors or different metabolic strategies, and those may depend on what is available in the environment. In some settings, nitrate or sulfate can substitute for oxygen, and the biofilm becomes a stratified chemical factory with distinct respiratory zones. In other settings, fermentation dominates, producing a distinctive chemical signature. In still others, the interior becomes a refuge of low activity. The geography of breath is, more generally, the geography of electron flow.

The biofilm matrix contributes to this geography in two ways at once.

Physically, it shapes diffusion paths and can slow the movement of molecules. Chemically, it can bind substances, concentrate them, or exclude them. EPS can trap enzymes that break down complex nutrients, effectively creating local food sources. It can also retain signaling molecules, amplifying communication within a neighborhood. But the matrix does not act uniformly. Stewart and colleagues in the 1990s emphasized that biofilms are heterogeneous even over distances of about ten micrometers: pores, channels, and density variations appear abruptly. That means neighboring pockets can experience very different oxygen histories, even if they are at the same "depth" in a crude sense.

This micron-scale heterogeneity helps explain a paradox that has long annoyed clinicians and engineers: why do biofilms respond so unevenly to treatments that, in principle, should penetrate everywhere? A disinfectant may reach the biofilm surface easily but be consumed or neutralized before it diffuses into the core. An antibiotic may diffuse in but encounter slow-growing cells that are intrinsically tolerant. A sudden change in oxygen availability may stress one zone while leaving another almost unaffected. The biofilm's response, from the outside, looks like stubbornness; from the inside, it looks like diversity.

A concrete example comes from chronic lung infections in cystic fibrosis, where thick mucus creates an environment with limited oxygen penetration. *Pseudomonas aeruginosa* can form biofilm-like aggregates within this mucus. In such settings, oxygen can be depleted rapidly, encouraging metabolic states that are less susceptible to certain antibiotics. The clinical details are complex and vary between patients, but the underlying principle is simple: structure plus consumption creates an oxygen gradient, and the gradient changes which physiological states dominate. The same organism can behave like different organisms depending on where it sits.

Another example, more domestic and less frightening, is the slimy layer in a sink drain or a showerhead. The outer layer is bathed in oxygenated water and experiences frequent disturbances. The interior, pressed against plastic or metal and protected by the matrix, can be oxygen-poor and slow-growing. When you scrub the surface, it comes off easily, but the regrowth can be rapid because the protected interior remnants remain. The persistence is not magic. It is geography.

The most arresting feature of this geography is how quickly it can shift. If flow increases, channels may deliver more oxygen, and zones that were fermentative can become respiratory. If a nutrient pulse arrives, oxygen may be consumed more rapidly, and zones that were respiratory can become oxygen-limited. If a portion of the biofilm sloughs off, buried cells can suddenly be exposed to oxygen and wake into growth. Biofilms are often described as stable, but their internal climates can be volatile. Stability at the city scale can coexist with turbulence at the neighborhood scale.

These shifts are not only responses; they can be strategies. A biofilm with layered physiology can be resilient in an uncertain world. If conditions favor rapid growth, the surface population exploits it. If conditions turn harsh, the slow interior persists. This is not necessarily altruism or planning; it is a natural consequence of spatial structure. Yet the outcome resembles a kind of distributed risk management. In a dangerous environment, a community that places all its metabolic bets on one state is fragile. A community that, by physics alone, maintains multiple states can endure.

None of this requires the cells to be different in principle. The same genetic program can produce multiple outcomes because gene expression—which genes are turned on and how strongly—depends on local conditions. A cell can sense oxygen directly, or sense redox state, or sense the byproducts of its neighbors, and adjust accordingly.

Over time, repeated exposure to certain microenvironments can even select for genetic changes that fit those conditions, but the immediate stratification happens faster than evolution. It happens within a single biofilm, within a single day, sometimes within hours.

A final twist is that oxygen gradients are not purely about scarcity; they are also about toxicity. Oxygen enables efficient energy extraction, but it also creates reactive oxygen species that can damage cells. Many bacteria maintain defenses against oxidative stress, and those defenses are energetically costly. In an anoxic interior, oxidative stress is lower, and some defenses can be relaxed. A biofilm thus contains zones that differ not only in energy yield but in the kinds of damage they must anticipate. The surface is a place of opportunity and hazard. The interior is a place of constraint and shelter.

Seen from outside, a biofilm can look like a uniform smear, a simple nuisance. Seen from within, it is a layered atmosphere: breathable boulevards, stuffy courtyards, sealed basements, and occasional drafty shafts where a channel brings fresh oxygen into the depths. The cells do not vote on where the air will be good. They build, consume, and adapt, and in doing so they create a geography in which breath is unevenly rationed, and life takes on different forms within the same microscopic city.

## 3.3 Slicing with Light

For much of microbiology's history, seeing meant killing. If a community was too thick to look through, it was scraped, smeared, stained, or sliced with a blade. These techniques are powerful and, in many settings, still indispensable. They can tell you which species are present, what molecules are being made, whether a cell is dividing or damaged. Yet they share a hidden cost: they turn a living, hydrated structure into a flattened record. A biofilm, in particular, does not forgive rough handling. Its matrix is water-rich, soft, and spatially intricate. Squash it, and the very feature that makes a biofilm a biofilm—its three-dimensional cityscape and its internal plumbing—vanishes.

The change arrived not as a new bacterium or a new drug, but as a new way of looking. Confocal laser scanning microscopy, usually shortened to *confocal microscopy* or *CLSM*, made it possible to image intact biofilms in three dimensions, in water, while they were still alive. It did not simply add prettier pictures to the literature; it changed what counted as a reasonable question. Once you could view a biofilm from the side without cutting it open, it became obvious that "thickness" was not a single number. Once you could capture stacks of images through depth and reconstruct them, it became possible to measure roughness, porosity, and surface area as quantitative properties rather than as metaphors. Once you could follow a biofilm over time without destroying it, it became possible to watch architecture emerge, not just infer it from the ruins.

The core idea of confocal microscopy is disarmingly simple: reject blur. In a standard wide-field microscope, the sample is illuminated and the objective collects light from a range of depths. If the sample is thin, that is fine. If it is thick—a hundred micrometers of hydrated biofilm, for example—light from above and below the focal plane floods the image. Details wash out into haze. The biofilm becomes a bright cloud with indistinct edges, and it is easy to convince yourself that it is more uniform than it truly is.

A confocal microscope adds a kind of gate. A laser illuminates a tiny spot in the sample, and a pinhole (or an equivalent optical trick) blocks most of the out-of-focus light on the way back to the detector. The result is an image of a thin "optical section": a slice through the sample that can be collected without physically cutting anything. By scanning the spot across the field of view, you build up a two-dimensional image. By stepping the focus up or down, you collect a stack of these slices through depth. Put the stack together and you have a three-dimensional volume, the microscopic equivalent of a CT scan.

The method matters in biofilms because thickness is common and transparency is deceptive. Biofilms can be highly hydrated and still densely populated. Under a bright-field microscope they may look like a faint film; under confocal they reveal towers, channels, and pockets of emptiness. Early work on optical sectioning of microbial biofilms by Lawrence and Korber (1991) helped establish that these communities are not simply compact layers but open, water-laden structures with species-specific patterns: some biofilms concentrate cells near the attachment surface, others build outward, and many contain large void spaces within what looks, from the outside, like a continuous mass. Those images were not merely illustrative; they were evidence that the city metaphor had literal architectural content.

Fluorescence made the revolution practical. Confocal microscopy works best when the sample emits light from specific structures rather than reflecting light from everything. Biofilm researchers began labeling cells and matrix with fluorescent dyes: molecules that glow when excited by the laser. Some stains bind DNA and mark cells. Others bind polysaccharides or proteins and highlight components of EPS. Genetic engineering added another approach: bacteria can be made to produce fluorescent proteins, turning them into living lanterns. With these tools, different cell types or physiological states can be colored differently, and the resulting images can show the spatial mosaic of the community in a way that black-and-white microscopy rarely can.

At the same time, fluorescence introduced new responsibilities. Light can be toxic. Strong illumination can bleach dyes and harm cells, especially during long time-lapse experiments. The confocal revolution therefore carried its own technical tension: the desire to watch biofilms grow and reorganize in real time versus the risk of changing what you are trying to observe. This tension did not invalidate confocal microscopy; it spurred refinements in instrumentation and experimental design, and it encouraged the development of alternative methods for certain questions.

One of the quiet achievements of confocal imaging was that it forced biofilm scientists to become more quantitative. A striking three-dimensional rendering can persuade, but it can also mislead. Two biofilms might look different simply because of different settings, different thresholds, different choices about how to color or smooth the volume. The field needed metrics—numbers that could be compared between experiments, strains, and labs.

That need led to computational tools that turned image stacks into measurements. COMSTAT, developed around the turn of the millennium and widely adopted in the early 2000s, became a standard way to quantify biofilm structure from confocal datasets. Instead of saying a biofilm looked "rough" or "thick," researchers could calculate mean thickness, biomass volume, surface-to-volume ratio, and roughness coefficients. These quantities are not merely descriptive. Surface-to-volume ratio, for example, is intimately tied to exchange with the environment: more surface relative to volume generally means more access to nutrients and oxygen, and more opportunity for waste to diffuse out. Roughness can influence how flow interacts with the surface and how pockets of shelter form. When architecture becomes a set of numbers, it becomes possible to link skyline to function in a disciplined way.

A historical anecdote captures how quickly this changed expectations. In the early biofilm literature, illustrations often showed a uniform layer of cells on a surface, a neat band of biomass with a boundary line. Those drawings were not dishonest; they reflected what could be seen with the tools at hand and what could be explained with the concepts available. Confocal imaging made those tidy cartoons feel quaint. It revealed that even "simple" laboratory biofilms often have dramatic local variation over distances of a few micrometers. A biofilm could be thick in one spot, thin in another; dense in one region, porous in the next; aerated along a channel, anoxic behind a ridge. Once you have seen those images, it becomes hard to believe that a single average thickness can capture the lived experience of the cells inside.

Confocal microscopy also changed how biofilm researchers interpreted earlier data. Consider an experiment measuring the overall rate of oxygen consumption by a biofilm. Without structural information, you might assume the entire community is participating similarly. With confocal images, you begin to suspect that most of the consumption is concentrated in a skin or in certain neighborhoods adjacent to channels. The rest may be dormant or running alternative chemistry. The same overall number can conceal very different internal realities, and confocal microscopy offered a way to connect global measurements to local behavior.

Real-time imaging made the story even more vivid. Instead of comparing early and late time points as separate samples, researchers could watch microcolonies grow, merge, and sometimes collapse. They could see channels open and close. They could see streamer-like filaments form under flow and whip in currents. They could observe that biofilm development is not always monotonic; parts can detach, leaving cavities, while other parts thicken. This dynamic view helped resolve a conceptual dispute about whether channels are "designed" features or accidental voids. When you watch channels persist and carry fluid, or watch them appear in reproducible patterns under certain flow regimes, it becomes harder to dismiss them as mere artifacts of preparation.

Yet confocal microscopy was never the only way to slice with light, and the last decade or two has broadened the imaging toolbox in ways that matter for biofilms. One line of development aims to reduce photobleaching and phototoxicity while keeping high resolution. Confocal light-sheet approaches, including single-objective designs that can be integrated with more standard microscopes, illuminate the sample in a way that can be gentler and faster for certain types of 3D imaging. Instead of scanning a single spot through an entire volume, they can illuminate a thin sheet (or an optical equivalent) and capture planes more efficiently. For living biofilms, where you might want to image repeatedly over hours, these methods help by limiting how much light each cell must endure.

Another line of development is almost the opposite in spirit: sacrifice some resolution in exchange for robustness and convenience. Optical coherence tomography (OCT) is a striking example. OCT is best known as a clinical imaging method, used routinely in ophthalmology to visualize layers in the retina. It is label-free, meaning it does not require dyes or genetic tags. It can image thicker structures than standard confocal microscopy, often in situ, and it can do so repeatedly with minimal disturbance. For biofilms growing in pipes, reactors, or other engineered systems where adding fluorescent labels is impractical, OCT offers a way to track thickness and volume over time and to connect those changes to operational conditions. It is less about single-cell detail and more about monitoring the city's skyline as it rises, erodes, and reshapes.

It is worth stressing what these tools do *not* do. They do not automatically tell you which molecules are present or which genes are active. Images can seduce: a bright tower looks like a thriving neighborhood, but it might contain many dead cells; a dark cavity looks empty, but it might be filled with matrix that does not stain. Confocal microscopy is most powerful when paired with additional information: stains that differentiate live and dead cells, reporters that indicate gene expression, sensors that map oxygen or pH, and experiments that perturb the environment. The image gives you the where; the additional tools help reveal the what and the why.

At the dinner-party level, the confocal revolution can be explained without optics jargon. A standard microscope looks at a thick object the way your eyes look into a fog: you see brightness but not depth. A confocal microscope looks at that same object the way a careful photographer might work in low visibility: by restricting what counts as the image and ignoring light that does not come from the plane of interest. Stack those planes, and the fog resolves into streets and buildings.

The practical implications have spilled far beyond basic science. In medical contexts, being able to image biofilms in three dimensions helps researchers understand why treatments fail and where to target them. In industrial settings, it helps engineers see how biofilms colonize surfaces, how they respond to changes in flow and nutrient supply, and how they contribute to fouling and corrosion. In environmental microbiology, it helps reveal the structure of microbial mats and the spatial relationships that enable complex cycles of carbon, nitrogen, and sulfur. A surprising amount of microbial ecology is architectural.

One concrete illustration of this architectural thinking comes from the study of mixed-species biofilms. A petri dish of mixed bacteria can look like chaos. Under confocal imaging with different fluorescent labels, that chaos can become organized: one species clustering near the surface, another in the depths; one forming a scaffold, another occupying channels; one dominating the cap of a mushroom-like structure while another persists at the base. These patterns are not mere curiosities. They influence who gets oxygen, who gets nutrients, who is exposed to stress, and who survives a disturbance. Without optical sectioning, it would be easy to mistake a mixed biofilm for a uniform soup.

The philosophical shift is easy to state: once biofilms became visible as three-dimensional structures, they became easier to think of as habitats rather than deposits. A deposit is something you remove. A habitat is something you must understand in context: the physics of transport, the chemistry of gradients, the mechanics of shear, the biology of growth and stress response. Confocal microscopy did not invent that complexity, but it made it difficult to ignore.

There is also a quiet emotional effect to these images that scientists do not always admit aloud. Confocal renderings of biofilms can be oddly beautiful: luminous ridges, vaulted cavities, branching channels like river deltas, towers that look uncannily deliberate. Beauty is not proof, but it can be a kind of motivation. It makes you linger. It makes you ask why a structure takes a particular form, and what the inhabitants experience within it. That curiosity, in turn, leads to experiments that connect form to function: to oxygen maps, to flow measurements, to gene expression patterns, to tolerance and persistence.

At the smallest scales, the biofilm city is built of cells and polymers. At the scale of human understanding, it is built of images and the tools that make them trustworthy. Slicing with light did not solve the mystery of biofilms so much as sharpen it into questions that can be measured: how steep is the oxygen cliff, exactly; how much of the interior is ventilated by channels; which neighborhoods are actively growing; how does architecture shift under stress; when does a tower become a streamer and a streamer become a clog. Under the laser's thin optical blade, the biofilm stops being a smear and becomes a place, with depth you can return to, slice by slice, until the patterns begin to feel less like slime and more like infrastructure.

## Chapter 4

# Chemical Conversations

*Far from being solitary mutes, bacteria inhabit a world rich in molecular chatter where chemical languages dictate war and peace. We uncover how these microscopic citizens conduct a census of their neighbors to coordinate bioluminescence or virulence, and how the dense architecture of the biofilm transforms into a bustling marketplace for trading genetic secrets.*

## 4.1 The Census Taker

A single bacterium is small enough to vanish in a raindrop, yet it can behave as if it has a plan. In a biofilm, that plan often looks coordinated: some cells produce sticky polymers that thicken the matrix, others dial up pumps that spit toxins back out, and still others conserve energy as if waiting out a siege. Coordination poses an awkward question. Bacteria have no eyes and no ears. They cannot look around and count their neighbors. How, then, does a cell decide whether it lives alone on an empty surface or in a crowd dense enough to justify expensive collective work?

One answer is that many microbes do not count bodies; they count chemistry. They conduct a census by sampling the concentration of small molecules in their surroundings, molecules they themselves release. When the molecules are scarce, the cell interprets that scarcity as loneliness. When the molecules accumulate beyond a threshold, it interprets the rising concentration as company. This logic is called *quorum sensing*: regulation of gene expression in response to changes in cell population density, mediated by secreted and detected signaling chemicals known as *autoinducers*.

A quorum is not a headcount; it is a turning point. The same cell, with the same DNA, can behave differently depending on whether its autoinducer "fog" is thin or thick. Below the threshold, the fog dissipates and nothing much happens. Above it, the fog triggers a switch in gene regulation that can alter metabolism, turn on virulence factors, reshape biofilm architecture, or—as in one of the most visually striking examples—make a bacterial population glow.

The threshold matters because so much of bacterial life involves costs paid now for benefits that only arrive if many neighbors pay too. Consider a secreted enzyme that breaks a big food molecule into smaller pieces. Making that enzyme costs energy and raw materials. If one cell makes it alone, the released enzyme may diffuse away, and any freed nutrients may be stolen by competitors. If thousands of cells make it together, the local environment changes: the enzyme concentration rises, digestion becomes efficient, and the group benefits. The same is true for many communal projects: releasing toxins, building extracellular polymeric substances (EPS) to fortify the biofilm, producing surfactants that alter surface tension, or emitting light that serves a host.

Quorum sensing is sometimes described as bacterial "talk," but the metaphor can mislead. A conversation implies intention and a shared understanding. Quorum sensing is more like a thermostat: a device that responds to the concentration of something in its environment and flips a circuit when a set point is reached. The cleverness lies not in conscious communication but in how evolution has wired chemical measurements to gene regulation.

At the molecular level, the logic is straightforward. Cells synthesize an autoinducer and release it. Because bacteria are tiny and molecules are restless, the autoinducer spreads by diffusion. At low cell density, most autoinducer molecules drift away into the larger world; any one cell, sampling its immediate neighborhood, detects very little. As the local population grows-on a tooth surface, in a wound, on a catheter-the same per-cell release leads to a rising background level. Now each cell detects more autoinducer, not because any one cell is emitting more, but because many are emitting together and the environment (especially the EPS-rich biofilm matrix) retains the molecules. When the concentration crosses a threshold, receptors and regulatory proteins inside the cells change state, and the population enters a new genetic program.

That program can be made sharper by a trick common in biology: positive feedback. Once a cell detects enough autoinducer, it increases the expression of the genes that make and detect that autoinducer. The system becomes self-amplifying. A few molecules past the threshold can lead to many more molecules, rapidly synchronizing the population. The result is a collective flip from one behavioral mode to another, a kind of microbial phase transition.

Immediately after grasping the threshold idea, it helps to see it. The point is not intricate biochemistry; it is the difference between signals that disperse and signals that accumulate, and how that difference can gate an on/off circuit like light production.

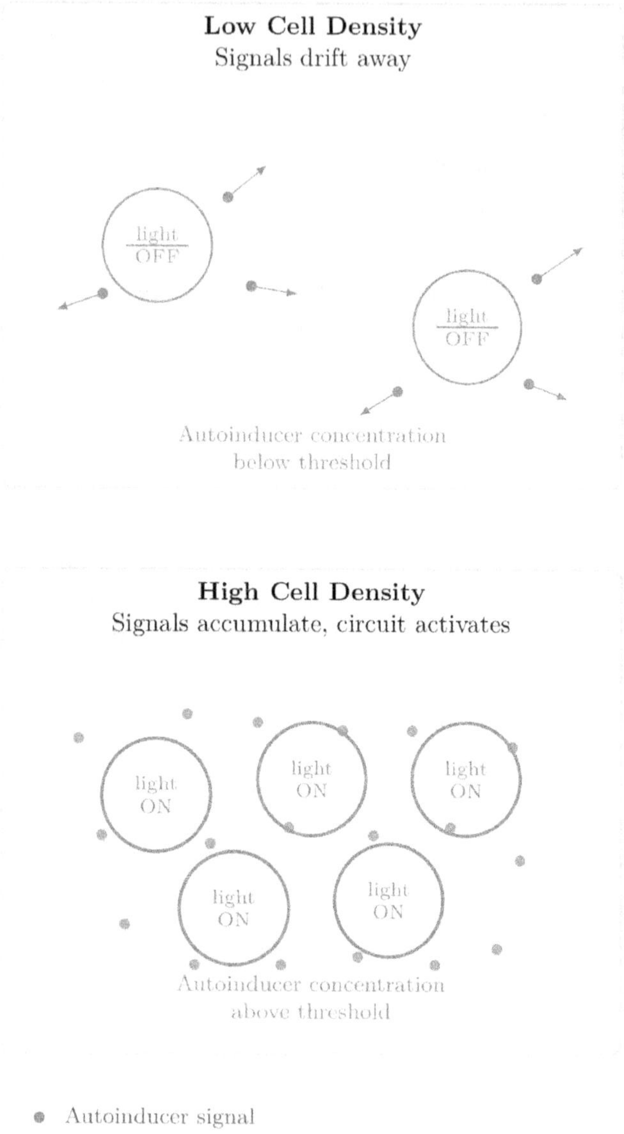


The best-known story of quorum sensing begins, appropriately, in the dark. In the late twentieth century, biologists studying marine symbiosis were captivated by the Hawaiian bobtail squid, a small nocturnal animal that hides from predators by erasing its shadow. It does this with help. The squid houses luminous bacteria-*Vibrio fischeri*-in a special light organ. When the squid swims under moonlight, it can match the brightness of the surface above, making itself harder for predators below to see. The bacteria get a protected home and a steady diet. The squid gets stealth.

The puzzle was that the bacteria did not glow all the time. Early in colonization, when only a few cells are present, there is little light. Later, when the population is dense, the light turns on strongly. A neat engineering-minded person might guess that the squid controls the bacteria like a dimmer switch. Instead, a major part of the control sits inside the bacteria: a density-dependent genetic circuit. The bacteria release an autoinducer; as they multiply in the confined space of the light organ, the signal accumulates; once it reaches a threshold, it binds to a regulatory protein that turns on the genes for luciferase and other components of bioluminescence. The result is a crowd-lit lantern, not a solitary spark.

Bonnie Bassler's work in the 1990s and early 2000s helped transform such observations into a general principle. Quorum sensing became recognized not as a marine oddity but as a widespread strategy: bacteria gauge local population density by measuring their own secreted signals, then coordinate group behaviors that only make sense in a crowd. The emphasis on *local* is crucial. A bacterium floating in open seawater can release a signal all day and never see it again. The same bacterium trapped in a crevice, a mucus layer, or the channels of a biofilm can end up bathing in its own molecules. Biology, like politics, depends on boundaries.

The boundaries of a biofilm are especially effective. The EPS matrix-sticky sugars, proteins, and DNA-turns a loose gathering of cells into a physical neighborhood. It slows diffusion, holds water, and creates tiny gradients of oxygen and nutrients. That same structure can trap autoinducers, allowing signals to build up at lower overall cell numbers than would be required in free fluid. A biofilm does not just provide a stage for quorum sensing; it can amplify it.

Different bacteria implement the census with different chemical currencies. Many Gram-negative bacteria use small molecules called acyl-homoserine lactones (AHLs), which can often slip across cell membranes. A typical arrangement is that one enzyme produces the AHL, the molecule diffuses out and back in, and inside the cell it binds to a transcription factor that then activates (or represses) specific genes. Many Gram-positive bacteria, by contrast, often use short peptides that are exported and then sensed by membrane-bound receptors, which trigger internal phosphorylation cascades. These details matter for pharmacology and for evolutionary history, but at the level of behavior they are variants on a shared logic: release, accumulate, detect, respond.

The response can be dramatic because quorum sensing is not limited to a single gene. Often it controls a *regulon*, a coordinated set of genes that shift the cell into a new mode. Think of it as switching a city from daytime routine to emergency operations. Streetlights come on. Police radios crackle. Factories shift production. The same buildings are present, but the priorities change.

In a biofilm, those priorities frequently involve defense. One reason biofilms are so hard to eradicate is that they are tolerant: not necessarily genetically resistant, but physiologically hard to kill. Antibiotics that devastate free-living bacteria can struggle against the slow-growing, matrix-protected, chemically buffered inhabitants of a biofilm. Quorum sensing can contribute to this tolerance by coordinating the production of matrix components, enzymes that neutralize threats, and collective behaviors that reshape the microenvironment. It is not the only driver-biofilm tolerance is multifactorial-but it can be part of the recipe.

A historical experiment from the 1990s made this concrete in a bacterium notorious in hospitals: *Pseudomonas aeruginosa.* This species forms robust biofilms and can cause chronic infections, especially in the lungs of people with cystic fibrosis and on medical devices. David Davies and Matthew Parsek (1998) examined mutants defective in producing a key quorum-sensing signal and found that, under their experimental conditions, the resulting biofilms were altered-flatter, less structured-and more susceptible to a detergent-like biocide. Adding back the signal could restore a more typical architecture. The message was not that quorum sensing is the sole architect of a biofilm, but that chemical census-taking can influence how a biofilm builds itself and how it withstands assault.

The idea that microbes can be induced to betray themselves by disrupting their signals has obvious appeal. If bacteria coordinate harmful behaviors-virulence factor secretion, toxin production, biofilm maturation-then perhaps sabotaging the signal could disarm them without killing them outright, reducing the evolutionary pressure for classical antibiotic resistance. This is the dream behind "quorum quenching," a term for strategies that degrade signals, block receptors, or otherwise jam communication. The dream remains compelling, but biology is rarely so obliging. Signals can be redundant. Pathways can compensate. Some signals serve multiple purposes. And bacteria are not passive targets; they evolve.

Even without therapeutics, quorum sensing forces a shift in how one imagines microbial life. It becomes difficult to maintain the picture of bacteria as purely individualistic. A better picture is conditional individuality: cells that behave alone when alone, and as a coordinated group when crowded. The conditionality is the key. It explains why a bacterium might "choose" (in the mechanistic sense) not to waste energy making a public good when there are too few neighbors to make the investment pay off.

That conditionality also explains why biofilms can change character so suddenly. Anyone who has watched a sink drain or a tooth surface go from clean to slimy knows the feeling of abruptness: nothing, nothing, and then—something. Part of that is just exponential growth. But quorum sensing adds a second kind of suddenness: a regulatory cliff. The population can grow smoothly while its behavior remains in one mode, and then, when the signal crosses a threshold, the group flips into another mode-one that may produce more EPS, alter motility, or switch on enzymes that remodel the surface. The physical texture of the biofilm can change quickly because the genetic program changes quickly.

It is tempting to treat the threshold as a fixed number, like a legal voting quorum: 51% and the motion passes. In practice, it is more like a dynamic standard that depends on the environment. How fast does the autoinducer diffuse away? Does the matrix trap it? Is fluid flow washing it out? Do neighboring species degrade it? Is the receptor sensitive? Is the cell's metabolism capable of producing the signal under current nutrient conditions? A biofilm on a catheter in a fast-flowing bloodstream will face a different signaling landscape than a biofilm in the relatively still fluid of dental plaque. A quorum, in other words, is not just about population density; it is about geography and physics.

That helps explain why quorum sensing is sometimes better understood as sensing *efficiency* rather than raw numbers. A cell that detects high autoinducer may infer, "I am in a place where secretions will stay local," which usually correlates with high density but can also reflect confinement. Conversely, a cell in a turbulent environment might "feel alone" even when many neighbors exist, because the molecules are carried away before they can accumulate. The census is always taken through the distortions of the medium.

The most profound effect of this chemical census might be psychological-ours, not the bacteria's. Humans like to divide life into solitary organisms and social ones. Ants coordinate. Wolves hunt in packs. Birds flock. Bacteria, we once assumed, simply multiply. Quorum sensing smudges that boundary. A microbial community can behave as a distributed organism, not by sharing a brain but by sharing a chemical environment. Each cell reads the same environmental signal, and because each cell carries a similar genetic circuit, the reading produces a coherent action.

The coherence can be eerily precise. In the squid symbiosis, the light is not a vague glow but a biologically useful output tuned to the host's needs. In infections, the output can be devastating: synchronized secretion of tissue-damaging enzymes or toxins, timed to occur when enough bacteria are present to overwhelm local defenses. In environmental biofilms, the output can be constructive: enzymes that break down complex polymers, coordinated to make the most of scarce resources. The same logic that powers a stealthy squid can power a stubborn chronic infection, because the logic is fundamentally about when it is worth acting together.

For all its elegance, quorum sensing does not imply harmony. A census can be taken for selfish reasons. A cell might use crowd information to exploit neighbors: to wait until others have invested in public goods and then freeload, or to time the release of toxins when competitors are vulnerable. Even within a single species, mutants can arise that benefit from the group without paying the costs. That tension-between collective benefit and individual advantage-plays out in microbial communities as surely as in human ones. Quorum sensing is one way evolution has managed the tension: by linking costly behaviors to conditions where the payoff is more likely.

Stand close enough to any busy human city and you can sense its population without counting: the density of voices, the smell of food stalls, the thrum of engines through pavement. A biofilm is also a city of sorts, and its inhabitants also take a census, not with eyes but with receptors, not with numbers but with molecules. The measurement is humble-just the concentration of a chemical in water-yet the consequences can be architectural. A thin scattering of cells behaves one way; a crowded neighborhood behaves another. In that difference lies much of the biofilm's power: its ability to turn proximity into collective action, and chemistry into a decision.

## 4.2 Esperanto of the Microbes

The first time you learn that bacteria can take a census with chemicals, it is natural to picture a private vote: a species counting its own kind, deciding when to flip from solitary thrift to collective expenditure. Real biofilms are rarely so tidy. A mature plaque biofilm on a tooth, a film on a river rock, a slimy layer inside a drain, or a community coating a catheter is less like a single-species town meeting and more like a crowded market where many accents mingle. The inhabitants jostle for space, trade metabolites, poison rivals, and sometimes cooperate with strangers. In that market, it helps to know not only "How many of *us* are here?" but "How busy is the neighborhood overall?" and, more pointedly, "Who else is listening?"

Microbes have evolved signals that behave like private dialects, but they have also evolved something closer to a shared tongue: chemical cues that can, under the right circumstances, be read across species boundaries. The most famous candidate for this role is a family of molecules loosely grouped under the name *Autoinducer-2*, often abbreviated AI-2. In the early years of quorum sensing research, AI-2 was sometimes described with the excitement reserved for a universal translator: a molecule produced by many species and sensed by many species, allowing a mixed community to coordinate as a whole.

The word "universal," though, is where the drama lies. A language is not only a sound or a script; it is an agreement between speakers and listeners. The same noise can be meaningless if no one has the right ears. The same chemical can be mere metabolic exhaust if no one has the right receptor. AI-2 sits in that ambiguous territory, and the ambiguity is instructive. It reveals how easy it is to mistake chemical coincidence for communication, and how much a biofilm depends on who can interpret what.

The story begins with a gene called *luxS*. In the late 1990s and early 2000s, as Bonnie Bassler and others were building a general theory of bacterial census-taking, *luxS* drew attention because it appeared across an impressive range of bacterial genomes. The gene is involved in making a small molecule derived from a metabolic intermediate. Many bacteria, it seemed, had the machinery to produce a potential signal that could leak into the environment. If AHLs were often "private" signals (common in certain groups but not all), then AI-2 looked like it might be the chemical equivalent of a shared marketplace language.

The tempting leap was: widespread production implies widespread conversation. The more careful reading is: widespread production implies widespread availability, but not necessarily widespread interpretation.

The molecule most often linked to *luxS* is not a single, stable chemical with a tidy structure that can be bottled and labeled. LuxS produces a compound called DPD (4,5-dihydroxy-2,3-pentanedione), which is chemically reactive and can rearrange itself in water into a mixture of related forms. Depending on the presence of boron and other environmental factors, different cyclic variants can dominate. "AI-2" is therefore a practical umbrella term: a set of interconverting molecules that can, in some organisms, trigger a quorum-sensing response. If that sounds messy, it is because chemistry in water often is. Cells thrive in soups, and soups rarely hold still.

This chemical shapeshifting matters because receptors recognize shapes. In a human port city, a phrase that means "hello" in one language may be nonsense in another; in biology, a molecule that fits one receptor may not fit another. A universal signal is not universal simply because many species excrete it. It is universal only if many species have evolved the capacity to detect it and respond in a way that matters.

Here the narrative takes a revealing turn. Dedicated AI-2 sensing systems are known, but they are not evenly distributed across bacteria. One well-studied pathway sits in vibrios, marine bacteria that include both luminous symbionts and pathogens. In these organisms, AI-2 can be detected by a receptor arrangement often described as a two-component system: a periplasmic binding protein that catches the molecule and a membrane sensor that relays the information inward through phosphorylation. Another pathway, different in logic, is found in some enteric bacteria: instead of sensing AI-2 at the surface and signaling inward, they import it via a transporter (the Lsr system) and then chemically modify it inside the cell. In that case, AI-2 becomes not only a cue but a kind of captured commodity, removed from the shared environment and converted into an internal signal state.

Those details are more than footnotes. They suggest two distinct social strategies. Surface sensing is like listening in the street: the chemical concentration outside becomes a measure of the local crowd. Uptake is more like collecting flyers and shredding them: the cell both reads the information and alters the information landscape for others. A species that imports AI-2 can, in effect, silence the public announcement system for its neighbors.

The distribution of receptors has fueled a long-running controversy.

Genomic surveys in the late 2000s argued that while *luxS* is widespread, unambiguous AI-2 receptor systems are much less so. The implication is sobering: many bacteria may make DPD primarily because they need LuxS for metabolism, not because they are "speaking" AI-2 in a social sense. In those organisms, AI-2 could be a metabolic byproduct that spills into the environment, chemically available but socially meaningless.

This is not a defeat for the idea of interspecies signaling; it is an insistence on precision. Biofilms are packed with chemicals that could, in principle, be used as signals. A true signal requires an evolved receiver and an evolved response. Without those, a molecule is simply part of the chemical weather.

The metabolic entanglement of LuxS is central to the confusion. LuxS participates in the activated methyl cycle, a core metabolic pathway involved in handling methyl groups, the small chemical tags cells use in countless reactions. Knocking out *luxS* in a bacterium can change growth, stress responses, and many downstream behaviors even if the organism has no AI-2 receptor at all. Early studies sometimes attributed the altered behavior of *luxS* mutants to "loss of AI-2 signaling" when, in fact, the mutant might be metabolically sick. This is the kind of experimental pitfall that haunts biology: remove a gene that does two jobs, and you may not know which job you disrupted.

The more mature view of AI-2 is therefore conditional: in some organisms, in some settings, AI-2 functions as a genuine communication system; in others, it is simply a molecule that happens to be there. That conditionality, far from diminishing the interest, makes AI-2 a perfect lens for thinking about multispecies life. A biofilm is not one conversation. It is many overlapping channels, some of which are intentionally received and some of which are merely overheard.

To see what "overheard" can mean, consider the difference between a deliberate message and an informative cue. In a café, the clatter of plates is not a message, but it tells you the place is busy. In microbial communities, many metabolites can act as cues of crowding or competition without being designed as signals. AI-2 may sometimes function this way: even if a bacterium does not have a classic AI-2 receptor, changes in the chemical environment associated with LuxS activity could correlate with community density or metabolic state. Other species with receptors might then treat AI-2 as a useful proxy for "how much bacterial life is nearby," whether or not the producers intended to communicate. Evolution is opportunistic; it can turn reliable byproducts into information.

Yet there are cases where the intentionality—again, in the evolutionary sense—is harder to dismiss. The most convincing examples come from organisms that clearly possess dedicated receptors and show coordinated gene regulation in response to AI-2. Here the shared tongue looks real: the species produces AI-2, releases it, detects it, and alters transcription in ways consistent with a density-dependent program. In mixed communities where multiple producers and multiple listeners coexist, AI-2 can plausibly provide a measure of total bacterial activity, not just one species' headcount.

Even then, "Esperanto" may not be the right metaphor. Esperanto was designed to be easy and neutral. Microbial shared signals are neither. They are embedded in competition and exploitation. They can be jammed, mimicked, or scavenged. A better metaphor might be a commercial pidgin: a limited shared vocabulary used for practical transactions, learned by some, ignored by others, and subject to cheating.

A concrete example of microbial multilingualism comes from a close relative of *Vibrio fischeri*: *Vibrio harveyi*. Work from the early 2000s showed that this bacterium does not rely on a single signal but integrates multiple autoinducers. One of these is species-specific, another (AI-2) is more widely produced, and yet another carries information about vibrios more broadly. The signals converge on a shared regulatory pathway that controls outputs including bioluminescence. The logic resembles a decision made from several sensors: "Are my close kin here?" "Are other vibrios here?" "Is the broader bacterial crowd here?" The answer is not a simple yes or no but a weighted integration that can tune behavior.

This matters for biofilms because biofilms are information-rich environments. The EPS matrix traps not only signals but also nutrients, toxins, and enzymes. Microenvironments form over millimeters: an oxygenated surface, an anoxic interior, acidic pockets, alkaline pockets. Signals diffuse through these gradients and are interpreted differently by cells in different niches. A shared cue like AI-2 can then act as a kind of background chatter, a community-level hum that is present across the structure, while species-specific signals provide local, private coordination. The result is not one choir singing in unison but several choirs sharing a building, occasionally aligning on a note.

The social consequences can be counterintuitive. A shared signal might encourage cooperation across species—say, coordinating the production of an extracellular matrix that benefits multiple residents. But it might also intensify competition, prompting bacteria to arm themselves when they detect a crowded neighborhood. In human cities, a dense crowd can mean opportunity or threat; in biofilms, it can mean both at once.

If AI-2 sometimes functions as a community signal, why would bacteria ever produce a public announcement that competitors can hear? The answer lies in the same economic logic that underlies quorum sensing more generally: a signal can benefit the producer if it helps synchronize behaviors among close relatives, even if eavesdroppers exist. Many bacteria in a biofilm are clonally related at small spatial scales; microcolonies often arise from one or a few founder cells. In that context, a "public" molecule can still be primarily a within-family coordination tool, because the local receivers are mostly kin. Competitors may hear the message too, but the net benefit to kin can outweigh the cost of being overheard.

Another possibility is that the signal's "publicness" is an unavoidable consequence of physics. Small molecules diffuse. Cells leak. If a metabolic intermediate like DPD is routinely produced, and if it reliably accumulates in crowded microenvironments, then it will inevitably become a cue that other organisms can exploit. The producer does not have to "want" to communicate for communication to evolve around it. Evolution can turn the environment itself into an information channel.

The ambiguity is not merely academic. It matters to medicine and public health because some chronic infections involve mixed communities. In the lungs of people with cystic fibrosis, in chronic wounds, and in dental plaque, multiple species coexist. Therapies aimed at disrupting one species' private quorum sensing may leave community-level cues intact. Conversely, therapies that target shared cues could have broader effects but also more unintended consequences, potentially altering beneficial communities. A molecule that looks like a villain in one setting might be part of a stable ecology in another.

The idea of a shared microbial language also forces a different way of imagining cause and effect in a biofilm. When a community shifts—becomes thicker, more tolerant to antibiotics, more inflammatory to the host—it is tempting to look for a single mastermind species, a microbial "mayor" giving orders. Shared signaling suggests a different model: distributed influence. One species can raise the ambient level of a cue, another can be particularly sensitive to it, and a third can remove it by uptake. The community phenotype can emerge from these interactions without any one species controlling the whole.

Picture, for example, a mixed biofilm where one bacterium produces AI-2 abundantly because its metabolism is running at full speed. Another species, equipped with an AI-2 detection system, interprets the rising level as a crowded neighborhood and turns on genes that thicken the matrix. A third species imports AI-2 via an uptake system, dampening the signal and thereby preventing matrix overproduction. None of these actions requires conscious coordination. Yet the result can look coordinated: the biofilm thickens or thins, becomes more or less permeable, more or less hospitable to newcomers. Chemistry becomes governance.

There is a human temptation to romanticize communication as cooperation. In microbial communities, communication is better understood as *information*. Information can be used to help, to harm, or simply to endure. A bacterium that can read a shared cue has an advantage: it can time its investments, avoid waste, and prepare defenses. In a world where antibiotics, immune systems, and rival microbes impose constant pressure, timing is survival.

The most honest conclusion about AI-2 and other purported universal signals is that microbial "Esperanto" is partial and situational. Some species speak it fluently, some only recognize a few phrases, and many merely generate the sounds as metabolic noise. That partiality does not make the language irrelevant. On the contrary, partial shared understanding is often what makes crowded places so dynamic. In a market where not everyone shares a native tongue, misunderstandings, approximations, and opportunistic translations can drive the very interactions that shape the community.

A biofilm, then, is not only a structure built from cells and glue; it is also a field of interpretable molecules. Some molecules are private passwords, understood only by close kin. Some are public cues, listened to by a wider set of neighbors. Some are simply the scent of life itself, drifting from metabolism into the shared water. The city's chemistry is never silent. Even when no one is "speaking," the air is full of information for any organism that has evolved the means to listen.

## 4.3 The Internet of Genes

A biofilm is a city built from cells and glue, but it is also a marketplace for information of a deeper kind. Chemical signals coordinate behavior; shared cues hint at who is nearby. There is, however, a more consequential form of traffic that moves through these crowded streets. It does not merely change what a microbe *does* today. It can change what it *is* tomorrow. Genes themselves can pass from one bacterium to another, sometimes across surprisingly wide evolutionary distances, and the tight physical intimacy of biofilms makes these exchanges more likely.

Genetic inheritance in animals and plants follows a familiar path: parent to offspring, generation by generation. Bacteria have that kind of inheritance too. When a cell divides, it copies its chromosome and gives each daughter a set. Yet bacteria also have a second mode of acquiring genes, often called *horizontal gene transfer* (HGT) because it moves information sideways, between neighbors rather than down a family tree. The metaphor that best captures the effect is not sex or marriage but software. A bacterium can acquire a new genetic "module" that alters what it can metabolize, how it resists toxins, or how it interacts with hosts, without waiting for slow mutation and selection. In the right conditions, that module can then spread through a community with unsettling speed.

Biofilms provide those conditions. Their defining feature is proximity: cells packed close, glued into stable neighborhoods. Proximity increases contact-dependent transfer. It also increases the time cells spend near the same partners, giving rare events more chances to occur. The matrix retains DNA, viruses, and other mobile elements, and protects fragile molecules from being washed away or degraded. Biofilms are not the only places HGT occurs, but they are particularly good at turning possibility into probability.

HGT comes in several classic forms, each with its own mechanics and its own poetry. *Transformation* is the uptake of free DNA from the environment: genetic litter scavenged and incorporated by a receptive cell. *Transduction* is gene transfer via bacterial viruses (phages), which can accidentally package host DNA and deliver it to new cells like a misaddressed parcel. *Conjugation* is transfer by direct contact, typically mediated by plasmids: small, often circular pieces of DNA that replicate independently of the chromosome and frequently carry genes that matter in crises, including antibiotic resistance.

Of these, conjugation best illustrates why biofilms are accelerators. It requires touch.

The mechanism has been described, a little mischievously, as "bacterial sex." The phrase is memorable, but it suggests something too symmetrical. Conjugation is more like a transaction. A donor cell carries a plasmid that encodes the machinery needed to transfer itself. That machinery includes a surface structure, often called a *pilus*, which can extend from the donor like a grappling hook. When it contacts a recipient, it brings the cells close, forms a bridge, and begins transferring a copy of the plasmid DNA. Once the recipient has the plasmid, it may become a donor in turn. The plasmid thus spreads, not because it is altruistic, but because it is a genetic element that has evolved to propagate.

The image is concrete enough to draw, and doing so helps keep the mind anchored in the physical reality. This is not an abstract exchange in a digital cloud. It is DNA moving through a microscopic bridge between two living cells.

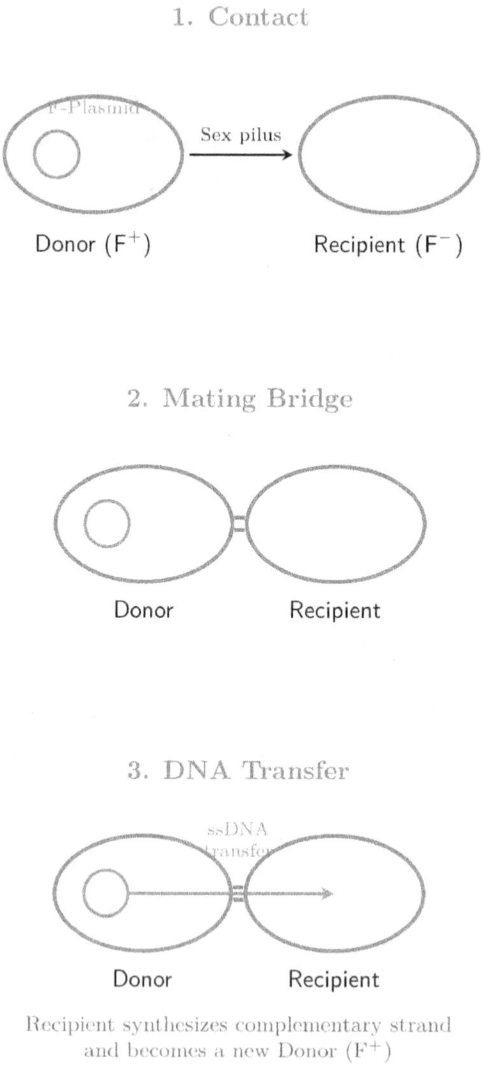

Once the physical steps are visible, the biofilm advantage becomes obvious. In a well-mixed liquid, donor and recipient may collide only briefly before drifting apart. In a biofilm, neighbors are not passing strangers; they are adjacent apartments. Contact structures need not hunt far. A pilus that finds a recipient is more likely to keep contact long enough for transfer. Dense packing also increases the number of potential recipients within reach. Conjugation is, in a literal sense, a neighborhood phenomenon.

The significance of conjugation is not that it happens-rare events happen everywhere-but that it can become common enough to matter. And when what is being transferred is antibiotic resistance, "matter" can mean a change in the outcome of an infection.

The history of medicine provides a vivid reason to take this seriously. The early antibiotic era in the mid-twentieth century brought genuine optimism: bacterial infections that once killed routinely could now be treated. Yet resistance followed swiftly. Some of that resistance came from mutation and selection within a lineage. Some came from the faster route: acquiring resistance genes from other bacteria. Plasmids carrying multiple resistance determinants became notorious. Hospitals, with their heavy antibiotic use and their abundance of surfaces—catheters, ventilators, sinks, drains—provided both selection pressure and biofilm-friendly habitats. The result was an environment where plasmids and the bacteria that carried them could thrive.

The fear is not merely that a bacterium becomes resistant, but that resistance becomes mobile. A gene that stays in one species is a problem; a gene that can hop is a crisis. Conjugative plasmids often carry genes that allow them to transfer across strains, and sometimes across species. Their host range varies: some are picky, others promiscuous. Biofilms, by mixing species in intimate contact, increase the odds that a plasmid will encounter a compatible recipient. The biofilm matrix also buffers cells from antibiotics, allowing donors and recipients to survive long enough for transfer to occur even under treatment.

Conjugation is not the only route by which biofilms trade genes. Transformation makes use of DNA that is already free. Biofilms are full of it. Cells die, burst, and release their chromosomes; some biofilms deliberately produce extracellular DNA as a structural component, a kind of molecular rebar that strengthens the matrix. That same extracellular DNA can become a genetic resource. Some bacteria enter a physiological state called *competence*, in which they actively take up DNA from their surroundings. The reasons for competence are still debated: it may provide nutrition, DNA repair templates, or new genes. Whatever the ultimate logic, competence turns the biofilm's DNA-rich environment into a library, albeit a library whose books are shredded and scattered.

Transduction, too, can be shaped by biofilm living. Phages move through microbial communities like predators and couriers combined. In a dense biofilm, phages may have many nearby hosts to infect, but the matrix can also hinder their diffusion, creating pockets where infections burn intensely and pockets where they barely reach. Some phages can carry bacterial genes from one host to the next, either accidentally or as part of specialized genetic programs. In a biofilm, where hosts are clustered, a phage that picks up a piece of DNA has a higher chance of delivering it quickly to another susceptible cell before the DNA is degraded or the phage is inactivated. The physical layout becomes part of the epidemiology.

The phrase "internet of genes" risks a misleading picture of frictionless exchange. Real gene transfer faces barriers, and bacteria are not naïve recipients. Cells have defenses that act like firewalls. Many bacteria cut foreign DNA using restriction enzymes, molecular scissors that recognize specific patterns and chop up unrecognized sequences. CRISPR systems provide a more adaptive memory of past phage attacks. Even if DNA enters the cell, it may fail to integrate into the chromosome if it is too different from the host's sequences. Plasmids, for their part, must be maintained: they need compatible replication machinery and must not impose too great a metabolic burden. A transferred gene can be useless or harmful in its new context. HGT is therefore not a free-for-all; it is a negotiated process shaped by compatibility and conflict.

Biofilms intensify both sides of this contest. They increase the exposure of cells to foreign DNA and mobile elements, but they also intensify selection for defense. In a dense, stable community, a successful phage can cause a local catastrophe, and a successful plasmid can reshape the genetic landscape of the neighborhood. The result is evolutionary pressure for both connectivity and security, a tension that mirrors our own digital lives.

The most sobering aspect of HGT in biofilms is that it can move traits that directly affect treatment. Antibiotic resistance is the headline, but it is not the only trait of concern. Genes that alter adhesion, secretion systems, toxin production, and metabolic flexibility can also spread. Even when a transferred gene does not become permanently fixed, it can still matter transiently. A plasmid that persists for weeks in a biofilm can change the community's behavior during an infection or in a hospital drain, even if it is later lost.

Newer work has suggested that cell-to-cell connections may be even more varied than the textbook trio of transformation, transduction, and conjugation implies. Reports of intercellular nanotubes, for example, describe thin membranous bridges that can connect cells and allow transfer of cytoplasmic materials, sometimes even between different species. The extent to which such structures contribute to gene flow in natural biofilms is still being mapped, and the field is cautious for good reason: extraordinary connectivity claims demand extraordinary evidence. Yet the very fact that researchers have observed such bridges at all is a reminder that dense communities encourage innovation in contact. When cells live wall-to-wall, membranes and molecules will find ways to interact.

The community perspective is also changing because measurement is changing. For decades, HGT was inferred from individual isolates and carefully controlled laboratory experiments. That approach taught the core mechanisms, but it struggled to capture the messy reality of mixed communities. More recent methods-metagenomics, which reads DNA from whole communities; proximity-ligation approaches that help link plasmids to their host cells; and clever reporter systems that light up when transfer occurs-are making it possible to ask which genes move, between whom, and under what conditions. One emerging theme is that gene transfer can be *punctuated*: not a constant trickle, but bursts triggered by stress, antibiotics, nutrient shifts, or other disturbances. A biofilm under attack may become, briefly, a particularly active market for genetic solutions.

This has an unsettling implication. Efforts to eradicate a biofilm can, in some circumstances, create the very conditions that favor the spread of mobile resistance: partial killing that leaves survivors under stress, high antibiotic concentrations that reward resistant variants, and dense communities that facilitate contact. It is a reminder that selection is always present, and that microbial communities respond to pressure not only by changing gene expression but by changing gene ownership.

None of this should be taken as an argument for despair. The same understanding that reveals the speed of gene exchange can guide better strategies: disrupting biofilm structure to reduce proximity, targeting plasmid maintenance, using phages thoughtfully, rotating therapies, and designing surfaces less hospitable to stable microbial neighborhoods. The point is not that bacteria are invincible, but that their evolutionary timescales and their modes of innovation differ from ours, and biofilms magnify those differences.

A final perspective helps keep the phenomenon in proportion. Genomes bear the scars and signatures of horizontal transfer; many bacterial lineages are mosaics assembled over time. Yet not every transfer event leads to an evolutionary breakthrough. Many acquisitions are transient passengers, lost when conditions change. Many genes arrive but never find a functional place in their new host. Adaptive transfers are a subset, and they become visible because they solve a problem. In biofilms, problems are abundant: toxins, starvation, immune assaults, antibiotics, competitors. That abundance makes the occasional useful gene more likely to be retained and spread.

In a dense microbial city, the most valuable commodity is often not food or space but capability. The ability to neutralize a drug, to cling more tightly, to exploit a new nutrient, to withstand a host defense: these can be encoded in short stretches of DNA, compact enough to travel. When cells are pressed together in the matrix, and when survival depends on rapid improvisation, genes become messages that can be passed hand to hand, bridge to bridge, until the neighborhood has changed in ways a lone bacterium could never have achieved on its own.

# Chapter 5

# The Politics of Slime

*Biofilms act as sophisticated micro-economies where individual survival frequently depends on collective investment. This chapter uncovers the fragile social contract binding these communities, revealing how they navigate the tension between altruistic cooperation and the inevitable rise of selfish exploitation.*

## 5.1 The Economy of Public Goods

A biofilm looks, at first glance, like a microbial accident: a smear on a rock, a film on a tooth, a stubborn residue in a pipe. Up close it is more like a crowded city built out of chemistry. Cells live pressed against one another, divided by microscopic streets of water, anchored to a surface, and wrapped in a shared gel. Life in that city is not sustained by individual self-sufficiency. It runs on shared infrastructure: molecules made by some cells that change the possibilities for everyone nearby. In human terms, these are the public works of the biofilm world: the roads, sewers, soup kitchens, and emergency services of a community that is too dense for solitary living.

Biologists use the phrase *public goods* for these shared resources. The term is borrowed from economics, where it refers to things that are costly to build and hard to restrict to paying customers: a lighthouse beam, a flood barrier, a streetlight. Microbial public goods are similar in spirit. They are products released from the cell that help with getting food, capturing scarce nutrients, building shelter, or surviving attacks. One cell can pay the cost of making them; many cells can enjoy the benefits.

This is easiest to understand if you picture a single bacterium trying to eat something it cannot swallow. Much of the world's carbon is locked up in polymers: starches, proteins, complex plant materials, dead tissue. A bacterium is small; it cannot engulf a chunk and digest it inside. So many bacteria digest outside their bodies. They secrete enzymes into their surroundings, enzymes that chop big molecules into bite-sized fragments. That act is generous and dangerous. Generous, because once an enzyme is out in the environment it does not carry a name tag. The sugars and amino acids it liberates can diffuse away. Dangerous, because producing enzymes is metabolically expensive: the cell must spend energy and raw materials to build large proteins, export them, and keep doing so as they wash away or break down.

A biofilm makes this strategy viable. In open water, secreted enzymes and their products are quickly diluted. On a surface, in a dense community, diffusion is slower and distances are shorter. The food fragments liberated by one cell's enzymes are more likely to be eaten by its near neighbours. That is already a hint of the biofilm's political economy: physical structure changes who benefits from generosity.

A second classic example is iron, the metal every organism needs for essential chemistry but which is often maddeningly unavailable. In oxygen-rich environments, iron tends to form insoluble compounds; in animals, it is locked away by proteins. Many bacteria solve this by secreting *siderophores*: small, high-affinity iron-binding molecules that roam the environment, grab iron, and return to cells via specific receptors. The trick is that siderophores are released into shared space. If a siderophore binds iron and drifts back through a neighbourhood, any cell with the right receptor can potentially take up the iron-loaded complex. The neighbourhood gains iron access, but the producers foot the bill.

One can hear the accountant's sigh in the phrase *metabolic cost*. Making public goods is a kind of tax: a diversion of finite budget away from immediate growth. Bacteria do not have money, but they do have something close enough to make the analogy useful. They have ATP (the molecular currency of energy), a supply of carbon and nitrogen skeletons for building biomass, and limited cellular machinery for transcription and translation. When a cell allocates ribosomes to making extracellular enzymes, it is not using those ribosomes to make the proteins that would let it divide faster. When it diverts carbon into sticky polysaccharides, it is not using that carbon to build new cell wall or DNA. In rich laboratory broth, the costs can look small; in the hard arithmetic of real habitats, they can be decisive.

The most conspicuous of these communal investments is the matrix itself: the extracellular polymeric substances, often abbreviated to EPS, that turn a crowd of cells into a cohesive film. The matrix is not a single substance but a composite: polysaccharides, proteins, lipids, and extracellular DNA woven together with water and ions. Calling it *slime* is accurate in texture but misleading in function. Slime suggests waste. The matrix is more like concrete and rebar mixed with insulation and glue, with a surprising amount of clever design. It lets cells adhere to a surface, stick to one another, and resist being swept away. It can trap nutrients, hold enzymes in place, and create microenvironments with different oxygen levels and pH. It can slow the penetration of toxins and shape how antibiotics reach their targets.

The matrix also behaves like a social budget because it is, in part, paid for collectively. The benefit is shared. A cell that makes sticky polymer helps stabilise the neighbourhood. A neighbour that contributes little may still enjoy the shelter of the communal gel. That does not make microbes moral, but it does make their lives political in the strict sense: their fitness depends on rules of association, on who pays and who benefits, and on how the group manages common resources.

A moment from the history of microbiology helps bring this into focus. In the late nineteenth century, as bacteriology was becoming a tool of medicine and industry, clinicians and laboratory workers were vexed by bacteria that formed ropy, viscous growths. Some pathogens produced thick capsules or slimy colonies that seemed to resist the host's defences. Capsule stains became a staple of the new microscopy, revealing halos around cells that were otherwise hard to see. Those halos were often interpreted as armour: a private coating that protected a cell from being eaten by immune cells. That is sometimes true. But when later researchers began to think seriously about surface-attached communities, it became clear that much of microbial slime is not merely armour for individuals. It is architecture for collectives. The same materials that make a capsule can also make bridges between cells, fill gaps, and create a shared physical space.

What makes this an *economy* rather than just a pile of useful secretions is that production is regulated and conditional. A city that poured all its budget into roads would starve its hospitals and schools. A bacterium that poured all its resources into matrix would stop dividing and be outcompeted. Biofilms therefore tend to produce public goods in ways that are responsive to circumstance. Some products are made only when cells sense they are in dense company, or only when nutrients are scarce, or only when the surface is stable enough that investment will not be washed away tomorrow.

A key concept here is the *range* of a public good: how far its benefits spread. In textbook economics, a public good is defined partly by non-excludability, the difficulty of keeping non-payers out. In the microbial world, excludability is often a matter of physics. How fast does the molecule diffuse? Does it stick to the matrix? Is it swept away by flow? Does it break down quickly? Even a very generous secretion can behave like a local club benefit if it barely travels beyond the producing lineage. Conversely, a molecule that diffuses freely and remains stable can become a truly global public good, a kind of open buffet for the community.

Consider secreted enzymes again. An enzyme that binds to the matrix, or works best right at the surface of the producing cell, will preferentially feed its maker and its closest relatives. An enzyme that floats freely in a fast-moving stream may feed whoever happens to be downstream. Biofilms in pipes and medical devices live inside flow, and flow can be as politically important as genetics. It can turn cooperation into charity and charity into waste.

The same is true for siderophores. They are small and diffusible, yet they are not always equally public. Cells must possess specific receptors to import iron bound to a particular siderophore. That specificity can turn a public good into a selective one: shared among those with compatible uptake machinery, useless to others. In mixed-species communities, this becomes a subtle form of economic protectionism. A species may flood the neighbourhood with a siderophore that it can retrieve efficiently, while competitors lacking the receptor watch the iron pass by as if behind glass. The community's chemistry contains both generosity and exclusion.

Public goods are not limited to food and metals. Some bacteria release molecules that change the environment itself. They may excrete acids that dissolve minerals, releasing trapped nutrients. They may secrete biosurfactants that alter the wetness and surface tension of their habitat, helping cells spread along a surface or access hydrophobic food sources. They may produce extracellular DNA that, in addition to its structural role in the matrix, can bind cations and alter local chemistry. They may release protective enzymes that detoxify reactive oxygen species, or molecules that bind and neutralise antibiotics. Each of these acts can be a personal defence and a communal service at the same time.

The book-keeping becomes most interesting when public goods unlock resources that solitary cells cannot access. Imagine a bacterium sitting on a surface coated with a complex protein. Alone, it might secrete a small amount of protease, liberating a trickle of amino acids that quickly diffuse away. In a dense biofilm, multiple cells secrete proteases into a matrix that holds them in place, generating a local cloud of nutrients. The cloud supports higher population density, which in turn produces more enzymes, which further increases access to food. Economists would recognise this as increasing returns and positive feedback. The biofilm's success is not merely the sum of its members; it comes from a collective ability to change the local rules of scarcity.

Those positive feedbacks are part of what makes biofilms so hard to remove from engineered and medical settings. A clean surface with a few planktonic cells is one thing; a surface with a mature matrix and a communal metabolism is another. In dentistry, for example, dental plaque is not simply bacteria stuck to enamel. It is a structured, multi-species biofilm in which public goods help maintain the local ecosystem. Some species break down dietary sugars and produce acids; others tolerate or even prefer those acidic conditions; still others consume metabolic by-products. Matrix components help retain acids and nutrients near the tooth surface, which is precisely what makes plaque a driver of tooth decay. The shared infrastructure is not morally neutral: it can be an engine of disease when it stabilises an unhealthy local chemistry.

The metabolic taxes that fund these public works vary widely. Producing a siderophore requires complex biosynthetic pathways. Producing a large extracellular enzyme requires significant protein synthesis and secretion machinery. Producing matrix polysaccharides can divert huge amounts of carbon. A striking feature of many biofilms is that they are *carbon-rich* relative to the cells embedded within them. The matrix can represent a substantial fraction of the biofilm's biomass. It is as if a city used an enormous amount of its resources not on citizens but on roads and buildings. That makes sense when the built environment is what allows citizens to survive floods, invasions, and famine.

Not all producers pay the same tax. In many species, matrix production is heterogeneous: some cells invest heavily, others less. This division of labour can arise from differences in local environment (oxygen-rich on the surface, oxygen-poor deep inside), from stochastic variation in gene expression, or from developmental programs that switch cells into different states. A biofilm can therefore contain something like specialized workers: cells that behave as builders, cells that behave as growers, cells that behave as dispersers ready to leave. The metaphor of a city becomes more than poetic when one realises that spatial position and cell state can map onto different economic roles.

Yet the word *public* can also mislead. Human public goods are ideally open to all citizens; microbial public goods often operate in gradients and neighbourhoods. A molecule released at one spot may never reach another. Benefits can be very local, and local benefits change the logic of cooperation. If a cell's investment mostly helps its closest kin, cooperation becomes less risky. In a biofilm, the physical matrix and the limited mixing of cells can keep relatives near each other. That arrangement is not a sentimental family reunion; it is a mechanism that makes the metabolic tax worth paying. A dollar spent on a streetlight in your own block feels different from a dollar that might end up lighting a street in a rival city.

A concrete example makes the balance between cost and benefit vivid. *Pseudomonas aeruginosa*, a bacterium notorious in hospital infections, produces a variety of secreted factors, including proteases and siderophores, that help it grow in difficult environments. It also produces polysaccharides that contribute to biofilm structure. In laboratory settings, mutants can arise that fail to make certain costly secretions yet still grow well when surrounded by producers. These mutants are not necessarily weak; they may reproduce faster because they have cut their metabolic taxes. The fact that such mutants can thrive at all tells us something important about the producers: much of what they make is indeed accessible to neighbours. The city's services can be used by residents who did not help build them.

This creates a tension at the heart of the biofilm economy. Public goods can lift the whole community into a new level of productivity, but they also create opportunities for exploitation. Even without talking about cheaters yet, the mere possibility reshapes what counts as a good investment. If the benefits leak too far, producers may be outcompeted by neighbours who pay less. If the benefits are tightly local, producers can recoup their costs through preferential access. Between these extremes lies a rich landscape of strategies: goods that are partly public, goods that are effectively private, goods shared only within a receptor-defined club, and goods whose value depends on flow, diffusion, and density.

The matrix itself illustrates this spectrum. Some matrix components behave like broad public goods, stabilising the entire structure and providing protection to many. Others appear to yield producer-biased returns because they stay close to the producing cell or because they directly enhance the producer's ability to adhere and remain in favourable positions. The same chemical can function differently depending on physical context. A polysaccharide that is exploitable in a well-mixed environment can become self-favouring in a thick, structured biofilm where it does not diffuse far. In microbial politics, physics is policy.

An intuitive way to understand the taxes and the returns is to imagine a simple thought experiment. Take a lone bacterium and give it a surface with a stubborn nutrient source, like a protein film. If it spends nothing on public goods, it may starve while sitting on food it cannot access. If it spends everything on enzymes, it may release nutrients but fail to build the internal machinery needed to capitalise on them. Somewhere between these extremes lies a strategy that maximises growth. Now place that bacterium in a crowd of close relatives and competitors. If its secretions mostly feed its relatives, investing in enzymes becomes more attractive: the benefits stay in the family, increasing the success of the genetic program that produced the enzymes in the first place. If its secretions mostly feed unrelated neighbours, investing becomes risky: the benefits may be captured by others, leaving the producer poorer for its generosity. The ideal investment policy is not fixed; it depends on who is nearby and how far molecules travel.

A biofilm therefore has a kind of invisible zoning map drawn by diffusion and flow. At micrometre scales, different neighbourhoods can have different economies. Near the surface where oxygen is abundant, cells may grow rapidly and produce enzymes that liberate nutrients. Deeper down, cells may face oxygen limitation and slow their metabolism, investing more in stress tolerance and matrix maintenance. At the edges, where the biofilm meets flowing water or host tissue, investment in adhesion and defence may be paramount. The public goods themselves help create these zones by shaping local chemistry: consuming oxygen, producing acids, binding metals, trapping nutrients.

There is something both unsettling and familiar about this. Human cities also create their own conditions. Infrastructure attracts people; people demand more infrastructure; the built environment changes the flow of resources and risks. A biofilm is a compressed version of the same logic, running on molecules rather than laws, but still shaped by trade-offs, spillovers, and neighbourhood effects.

A final twist is that the biofilm's public goods are not only about growth; they are also about security. The matrix can slow the entry of antibiotics and immune factors. Secreted enzymes can neutralise threats. Some secretions can even function as weapons, damaging rivals and protecting the producers' local monopoly on resources. In such cases, the line between a public good and a public defence blurs. A fortified wall benefits everyone inside, but it is built for a reason: there are enemies outside, and sometimes inside.

Thinking this way changes how one sees that ordinary smear on a surface. The slime is not just residue. It is investment. It is a transfer of private cellular wealth into shared infrastructure that reshapes the environment, making new forms of life possible in places where solitary cells would fail. A biofilm city is built molecule by molecule, paid for in energy and opportunity, and sustained because, in the right neighbourhood, the returns on cooperation can be higher than the returns on going it alone.

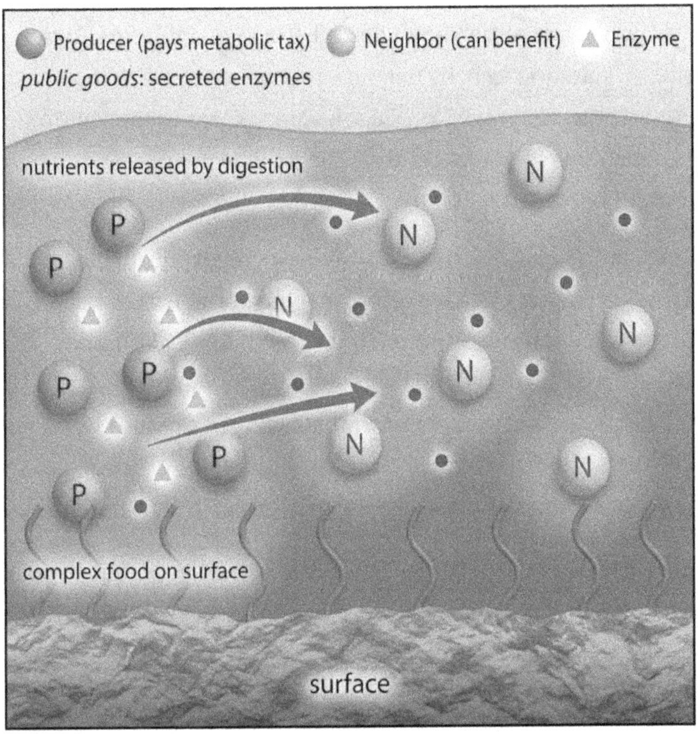


## 5.2 Invasion of the Free-Riders

The chemistry that makes a biofilm feel like a single organism also makes it vulnerable. Shared enzymes turn stubborn food into edible fragments. Shared iron-scavengers pull essential metals out of environments that would otherwise starve cells. Shared matrix holds the community in place when water or mucus tries to shear it off. These are triumphs of collective engineering, but they come with a weakness familiar from human life: once a benefit can be enjoyed by neighbours, a neighbour can enjoy it without paying for it.

A *cheater* (often called a *free-rider*) is a cell that takes advantage of communal resources while contributing less than its neighbours. The word sounds moral, but it describes a very ordinary evolutionary event. A mutation knocks out a costly behaviour. The mutant saves energy. If the communal product is still present in the environment, the mutant can grow faster than the producers around it. Natural selection, blind and local, tends to favour that faster growth.

The unsettling part is how easy it is for cheating to begin. One does not need a devious microbe, just the normal error rate of DNA copying. In a large biofilm, there are astronomical numbers of cell divisions. Every day, somewhere in the city, a mutation will appear that trims a metabolic expense. If that trimmed expense was a public good, and if the mutant remains close enough to benefit from its neighbours' contributions, a free-rider lineage has been born.

This is the microbial version of an argument that burst into public consciousness in the twentieth century. In 1968, the ecologist Garrett Hardin popularised the phrase *the tragedy of the commons.* He asked readers to imagine a shared pasture open to all. Each herder gains by adding one more animal, because the profit from that animal accrues to the individual, while the damage from overgrazing is spread across the whole group. Rational self-interest, repeated enough times, ruins the pasture for everyone. Hardin's essay was about human institutions, and it has been criticised for oversimplifying how real communities manage shared resources, but the core logic is brutally clean. Benefits that are individually captured and collectively funded invite overuse.

Biofilms stage this tragedy in miniature. The currency is not cattle but proteins, sugars, iron, and protection. The time scale is not decades but hours or days. The outcome can be as stark: a community that once grew robustly becomes thin, fragile, and slow, because too few cells are paying the taxes that keep the communal systems running.

The classic cheater story begins with a trait that is costly, shared, and not easily restricted to contributors. Secreted enzymes are an obvious target. If the neighbourhood is full of proteases that digest proteins into amino acids, then a mutant that stops making protease still eats amino acids. It gets the meal without the kitchen labour. Siderophores offer an even sharper illustration. Producing a siderophore requires a suite of specialised enzymes and a continuing drain of materials. If iron-loaded siderophores drift through the matrix, any cell with the receptor can import them. A mutant that retains the receptor but stops producing the siderophore is, in effect, importing iron on someone else's budget.

The dynamics can be counterintuitive. A cheater often thrives precisely because the community is healthy. When public goods are abundant, the cheater's savings matter most. It can grow and divide rapidly, increasing its frequency in the population. From a distance, the biofilm may still look fine; there are still many producers, and the goods are still present. Only later, when cheaters become common enough, does the trouble show. Public good production drops. Digestion slows. Iron becomes scarce again. The matrix thins. Growth falters for everyone, cheaters included, because the entire economy has been undermined.

That lag between cause and collapse is one reason cheating is so difficult to prevent. It resembles a civic failure that arrives long after the budget cuts: the potholes appear months later, the bridge fails years later, and the political argument has already moved on.

Quorum sensing provides a particularly vivid route into this problem. Quorum sensing is a way bacteria use chemical signals to estimate local density and coordinate behaviours that are more effective when performed together. In many species, a cell releases a small signalling molecule; as the population grows, the signal accumulates; when it crosses a threshold, it triggers changes in gene expression.

One of the most common outcomes is the activation of public goods: extracellular enzymes, toxins, surfactants, and factors involved in building or remodelling the matrix. From the point of view of a cooperative community, quorum sensing is common sense. Don't pay for an expensive public works project until there are enough taxpayers to make it worthwhile.

From the point of view of a potential cheater, quorum sensing can be an invitation. If public goods are turned on by a signal that diffuses through the biofilm, a mutant that fails to respond to the signal may avoid the cost of producing those goods while still living in a neighbourhood where others do respond. In *Pseudomonas aeruginosa*, a bacterium famous for forming biofilms in lungs, wounds, and medical devices, quorum sensing controls a suite of secreted products, including proteases and other factors that help the population exploit complex nutrients. Mutants in key regulators (often discussed in terms of *lasR* variants) can arise that contribute less to the communal pool while still enjoying its benefits.

The appeal of this strategy is immediate when one thinks in metabolic terms. Protein secretion is expensive. So is the synthesis of secondary metabolites. A mutant that stops doing these things reallocates resources to growth and division. In a rich communal environment, that shift can be enough to outpace its neighbours. The cheater is not necessarily more *aggressive*; it is simply more *economical* with its internal budget.

Clinically, the consequences can be complicated. A biofilm is not just a pile of cells; it is a protective structure. Many of the communal traits that help a biofilm persist against immune attacks and antibiotics are regulated, directly or indirectly, by these cooperative pathways. When cheating spreads, the biofilm may become less robust. A community full of free-riders can be thinner and less cohesive, and in some laboratory observations, mixed populations that include quorum-sensing cheats can show increased susceptibility to certain treatments because collective defences are weakened. The irony is sharp: a mutation that benefits a lineage in the short term can make the whole community, including that lineage, more vulnerable to external threats.

The logic of cheating also explains why the word *public* should be used carefully. Some goods are shared, but not shared equally. A molecule may be secreted, yet remain close enough to its producer that nonproducers cannot exploit it effectively. Some matrix components, for example, appear to behave less like open public goods and more like local club goods: they help neighbours, but they help producers disproportionately, especially in structured biofilms where lineages form clusters. Under those conditions, a nonproducing mutant cannot simply sit back and enjoy the benefits. It might receive a little shelter, but not enough to outgrow the producers that are building the shelter.

This distinction matters because it sets the stage for when cheating becomes a genuine invasion rather than a minor nuisance. A cheater can only spread if it gains more from the neighbourhood than it loses by withholding contribution. That balance depends on physics (how far goods diffuse), genetics (whether receptors are shared), and ecology (whether the population is clonal or mixed, whether flow washes goods away, whether nutrients are patchy). In a well-mixed soup, cheating is often easier, because goods are widely shared. In a structured biofilm, cheating may be harder, because benefits are local and often biased toward relatives.

Even when cheating is possible, it rarely produces a tidy morality play. Natural communities are messy. A cheater may be a cheater with respect to one public good but a contributor to another. A mutant that stops producing a siderophore might still produce matrix. A cell that is a parasite in one context could be a benefactor in another. Environments also change: flow rates shift, oxygen gradients steepen, nutrients arrive in pulses. The economics of contribution and exploitation can be rewritten hour by hour.

Still, certain patterns show up again and again, and they can be drawn with the simplicity of a classroom graph. One line tracks cheaters' frequency: low at first, then rising as they enjoy the public goods without paying. Another line tracks total population density: high and stable at first, because producers keep the system running, then falling as production collapses. The crucial feature is the delay. Cheater dominance comes first; system collapse follows. By the time the crash arrives, the cheaters have already won the local competition, only to discover they have inherited a poorer city.

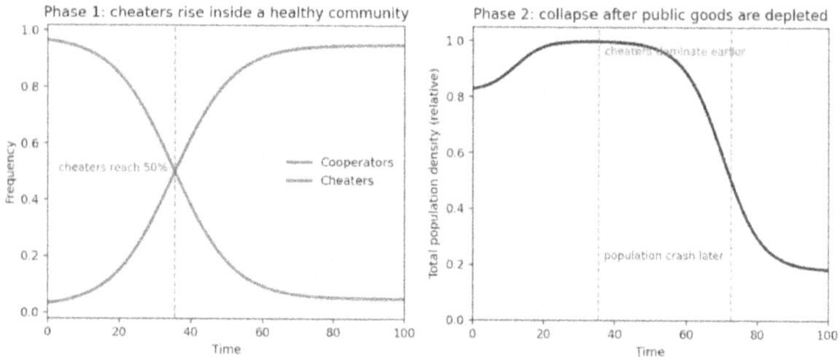


The figure is stylised, but the shape captures a biological reality: selection can reward short-term savings even when those savings erode the very conditions that made growth possible.

There is another, subtler danger in the invasion of free-riders. A community can become trapped in a low-productivity state even if it does not collapse entirely. Imagine a biofilm where enough public goods are produced to keep the population alive, but not enough to allow the robust growth seen when cooperators dominate. The result is a stagnant city: sparse, brittle, and more exposed to external hazards. In medicine, that might mean a biofilm that is easier to disrupt but still persistent enough to cause chronic infection. In industry, it might mean a fouling biofilm that repeatedly re-forms after cleaning because its remnants maintain a minimal matrix and a protected seed population.

Cheating can also change the biofilm's relationship with other species. When a community's shared resources are abundant, outsiders may be able to exploit them too. A neighbouring species with the right receptors might import siderophore-bound iron without producing the siderophore itself. A species tolerant of the matrix environment might move in and feed on digestion products. In this way, internal free-riding and external opportunism can reinforce each other: weakened cooperation within a species may open space for other species that further destabilise the original cooperative system.

Yet it would be a mistake to think of the cheater as an external invader in every case. Often it is a citizen born inside the city, genetically almost identical to its neighbours, differing by a mutation or two. That closeness is what makes the story so sharp. The enemy is not foreign; it is the ever-present possibility that a cheaper variant of the same life strategy will appear and take over.

A helpful way to hold all this in mind is to picture the biofilm as an agreement written in molecules rather than words. Producers obey the agreement by investing in shared goods. Cheaters break it by withholding investment. Evolution is the audit. If breaking the agreement leads to a higher individual payoff, cheaters increase. If breaking it destroys the value of the agreement itself, everyone loses. The tragedy is not a moral lesson delivered by nature; it is a description of what happens when selection operates at the level of individuals while benefits depend on collective investment.

There is also a kind of pathos in the microscopic scale of it. A single cell, mutating in the darkness of a biofilm, does not foresee a future where the communal protease pool runs low, where iron becomes inaccessible, where the matrix becomes porous. It only experiences a present where growth is possible, and where not paying for public goods makes that growth a little faster. Multiply that logic across thousands of generations and the city's prosperity is quietly converted into private savings, until there is nothing left to save. The end state is not dramatic heroism or villainy, but a biofilm that can no longer afford the chemistry that once made it strong, a community living on the remnants of its own dismantled infrastructure.

## 5.3 Law and Order in the Matrix

A city that relies on shared services cannot survive on goodwill alone. If the streetlights are paid for voluntarily, someone eventually notices they can read by their neighbour's lamp. If the fire brigade comes whether or not you paid the tax, someone eventually stops paying. In biofilms, the same arithmetic operates at molecular scale. Free-riders can spread because public goods are real and costly. That does not mean biofilms are doomed to collapse. It means that stable co-operation requires constraints: ways of keeping benefits close, tying contribution to survival, or making parasitism expensive.

The surprise, once one looks for it, is how many kinds of constraint microbes have at their disposal. Some are physical, built into the architecture of the matrix and the limits of diffusion. Some are genetic, built into regulatory wiring that couples public behaviour to private necessity. Some are outright punitive: biochemical sanctions that fall harder on cheaters than on cooperators. None of this requires intention. The biofilm does not *decide* to police; lineages that happen to make exploitation difficult are the lineages that persist.

The simplest form of law is geography. In a stirred liquid, a public good spreads widely. In a structured biofilm, it does not. A cell divides and its daughters stay close, pushing outward like bricks laid on top of earlier bricks. As the colony thickens, relatives form clusters. The shared matrix becomes a kind of neighbourhood boundary, and chemistry becomes local.

This locality turns the economics of public goods on its head. When enzymes, siderophores, or matrix components mostly benefit nearby cells, the beneficiaries are disproportionately kin. A cell that pays a metabolic tax is not donating to an anonymous city; it is subsidising its own genetic lineage in the next few micrometres. Social evolution theorists call this *kin selection*. The term can sound sentimental, but it is mostly a statement about proximity. Helping your relatives works when relatives are the ones you are likely to help. The matrix, by keeping families together and by slowing the travel of secreted molecules, can make that true.

Even very small changes in mixing can decide whether cheaters flourish. Picture two extremes. In the first, cells are constantly churned like grains in a shaken jar; every secreted product becomes widely shared. In the second, cells grow in place, with little rearrangement, and secreted products are mostly captured locally. The first condition invites free-riding because the benefits are easy to steal. The second makes stealing harder because the best access to the goods is within the producer's own cluster. Many biofilms sit somewhere between these extremes, but the principle is robust: the more structured the habitat, the more cooperation can pay.

A biofilm is not only structured by the stickiness of the matrix; it is structured by the movement of water. Flow is often imagined as the enemy of biofilms, a force that should rip cells away. In reality, flow can serve as a kind of economic regulation. If soluble public goods are washed away quickly, a cheater cannot sit at a distance and still benefit. The interaction radius shrinks. Producers that remain close to their own secretions are favoured, while distant exploiters are starved of the very goods they hoped to steal. Water, in this sense, acts like an auditor that limits how far communal wealth can leak.

Flow also creates a harsh kind of accountability. When digestion products are removed by moving fluid, the only way to benefit is to capture them quickly, often very near where they are produced. A cheater that saves energy by not producing enzymes still faces the same fast-moving stream; unless it sits right in the productive neighbourhood, it will be hungry. Biofilms that grow in pipes, streams, and on medical devices are constantly negotiating this balance between retention and loss. The matrix can hold cells and enzymes in place, but the environment keeps testing the strength of that grip.

Physical law is powerful, but it is not the only law. Genetics can hard-wire cooperation to private benefit in ways that make cheating self-defeating. One common strategy is *linkage*: putting public-good production under the control of regulatory systems that also control traits that a cell cannot do without. If the switch that turns on secreted enzymes also turns on essential metabolic functions, then a mutant that breaks the switch harms itself. It might avoid the cost of cooperation, but it also loses access to something required for growth or survival in that environment.

This coupling can happen in several ways. Sometimes the same regulator controls both public and private genes. Sometimes the metabolic pathway that produces a public good also produces intermediates needed internally. Sometimes the public good is made in a form that is inherently producer-biased: it sticks to the producer's surface, or it strengthens the producer's own attachment more than it strengthens a neighbour's. The details vary by species and by molecule, but the underlying principle is the same. Stable cooperation does not require altruism; it requires that contribution cannot be easily separated from self-interest.

A concrete example comes again from *Pseudomonas aeruginosa*, a bacterium whose social life has been dissected with unusual care because of its role in chronic infections. In laboratory biofilms, quorum sensing mutants can arise that contribute less to certain shared products, behaving as cheats. Yet *P. aeruginosa* also possesses mechanisms that can make cheating costly. One striking case described in the mid-2010s involved a form of policing linked to quorum sensing: cooperators produced a toxin (hydrogen cyanide) under quorum-sensing control, and the same regulatory network also influenced resistance to that toxin. The consequence is intuitively satisfying in a grim way. Cells that participate fully in the quorum-sensing programme are better prepared for the toxic environment that programme creates. Cheaters that fail to turn on the programme can be disproportionately harmed. Punishment is not administered by a judge; it is built into the chemistry of a shared regulatory state.

The lesson is not that bacteria are vindictive. It is that a regulatory cascade can create a private benefit for cooperators and a private cost for cheats, even when the trait under discussion is nominally public. One can imagine a city in which paying taxes automatically enrolls you in fire insurance while also funding the fire brigade. If you dodge the tax, you lose the insurance. In microbes, the same logic is executed with gene expression rather than paperwork.

Policing does not have to be purely within-species. In a mixed community, public goods can be exploited by other species, not just by mutant neighbours. This is especially relevant for products like proteases that liberate nutrients anyone can eat. Here, some bacteria respond by linking cooperation to aggression. *P. aeruginosa*, again, provides an illustrative case: quorum sensing can regulate both cooperative goods (like extracellular enzymes) and antagonistic factors (antimicrobials, toxins, secretion systems) that harm competitors.

The coupling makes evolutionary sense. If you are going to invest in digesting a food source that will feed anyone in the vicinity, it helps to ensure the vicinity is not full of freeloading strangers. A molecule that looks like a weapon in one context can function as a stabiliser of cooperation in another by reducing interspecies exploitation.

A biofilm is therefore not merely cooperative; it can be selectively cooperative, maintaining a shared economy among close relatives while excluding outsiders. Receptor specificity in siderophore uptake is one such exclusion mechanism: only cells with the right receptors can cash in the iron-bound cheque. Aggression is another. The biofilm matrix itself can contribute, because it is not just a passive gel. Its chemistry can favour certain cells and impede others. When matrix components bind enzymes or signals, they can create microenvironments where a cooperating lineage has privileged access. Outsiders may be physically present yet chemically poor.

These strategies are not foolproof. Cheats still appear. What changes is the balance of power: whether a cheater lineage can sweep through the city or remains a marginal population living on scraps.

Spatial structure deserves a closer look because it is so often the deciding factor, and because it can be described with a very plain picture. Imagine a well-mixed environment: cells and molecules are constantly stirred so that everyone meets everyone. In such a world, a cheater that does not produce a public good can still access it almost as easily as a producer. The cheater's advantage is mainly the saved cost, so cheating spreads readily. Now imagine a structured biofilm: cells grow in place, relatives cluster, and secreted products do not travel far. In such a world, producers mostly feed and protect their own cluster. A cheater that arises inside a cluster may gain locally, but it also weakens its own relatives. If it is surrounded by producers that are its close kin, its success does not translate into long-term lineage success unless the cluster as a whole remains productive. Many cheats are therefore punished not by an external police force, but by their own dependence on a local economy that collapses first around them.

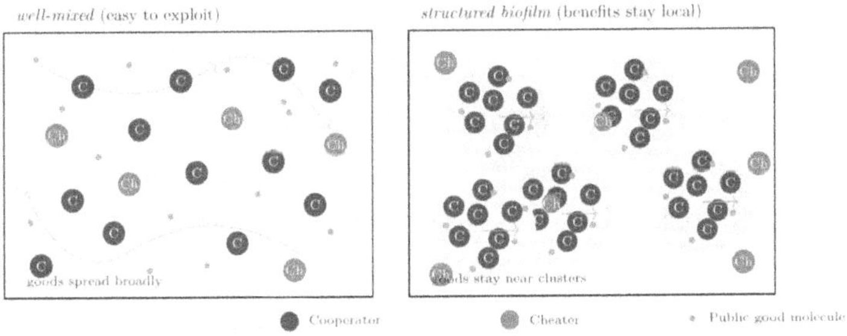


Spatial structure, though, is not a guarantee of harmony. It is more like the difference between a neighbourhood market and a city-wide free buffet. It narrows the scope of exploitation, but it does not eliminate the temptation. A cheater born inside a productive cluster can still siphon benefits, and if it becomes common within that cluster it can still damage the local economy. The biofilm's stability often comes from layering multiple mechanisms on top of one another: physical localisation plus genetic coupling plus selective punishment plus selective exclusion.

There is a further complication that feels uncomfortably familiar. Enforcement mechanisms can have costs too. A toxin that punishes cheats can also harm cooperators if the dosage is high or if conditions change. An antimicrobial that excludes other species may provoke arms races. Tight coupling between cooperation and private functions can reduce flexibility, making it harder for a population to adapt if the environment changes. Every law has unintended consequences, and in microbial communities those consequences are paid in growth rate and survival.

Biofilms also govern themselves over time. A mature biofilm is not a permanent structure; it is a lifecycle with phases of building, maintaining, and dispersing. Some bacteria remodel their matrix as conditions change, trimming or altering components to shift how tightly cells are bound. Work in *Vibrio cholerae*, the cholera bacterium, has shown that matrix components can be downregulated and enzymes can modify the matrix to facilitate dispersal. This is less like abandoning a city and more like opening the gates at a particular season, letting groups leave when the local environment becomes crowded or dangerous. The same regulatory logic that coordinates cooperation can also coordinate exit, and exit itself can act as a form of enforcement: lineages that cannot contribute to a stable matrix may find themselves excluded from the departing group, left behind in a depleted neighbourhood.

These ideas matter beyond intellectual satisfaction because they point to a different way of thinking about anti-biofilm strategies. Conventional antibiotics target essential private functions: cell wall synthesis, protein synthesis, DNA replication. That creates strong selection for resistance, because a resistant mutant gains an immediate survival advantage. Public goods are different targets. If one interferes with a cooperative trait, the evolutionary incentives can be less straightforward. A mutant that restores the cooperative trait might pay a cost that benefits its neighbours, including those still susceptible to the inhibitor. Under some conditions, this can lead to the counterintuitive outcome that resistance to an anti-cooperation strategy is disfavoured: resistant producers subsidise susceptible nonproducers. Experiments in organisms such as *Salmonella* have explored this logic by targeting matrix-related cooperative traits, suggesting that undermining shared infrastructure can sometimes be an evolutionarily robust way to weaken biofilms.

The promise should not be romanticised. Real infections and industrial biofilms are complex, multi-species, and embedded in host or engineered environments that change over time. Still, the underlying message is practical: if a biofilm behaves like a city, then interventions can aim not only to kill individual cells, but to destabilise the rules that keep the city prosperous. Sometimes the fastest way to empty a fortress is not to batter every stone, but to cut the supply lines and make the walls too expensive to maintain.

Law and order in the matrix is ultimately a story about constraint in a world of cheap mutation. Cooperation is always at risk because it produces exploitable surplus. Biofilms persist because their architecture and genetics make that surplus harder to steal, and because some lineages have evolved ways to make theft hurt. The result is not a utopia of perfect cooperation but a working, tense equilibrium: clusters of relatives investing in local public works, signalling and regulating when to spend, sometimes punishing defectors, often excluding outsiders, always negotiating with the physics of diffusion and flow. On a surface that looks smooth to the naked eye, an entire civic drama can play out in a space smaller than the width of a hair, with consequences that reach as far as a patient's lungs or a city's water supply.

## Chapter 6

# The Unconquerable Fortress

*Why do standard antibiotic courses often fail to clear chronic infections, even when laboratory tests suggest the bacteria are vulnerable? The answer lies not merely in genetic mutation, but in the physical and physiological architecture of the biofilm city itself. We uncover the multi-layered defense strategy-from chemical sponges to sleeper cells-that transforms a colony into a bunker capable of withstanding chemical monsoons and biological sieges.*

## 6.1 The Chemical Sponge

A biofilm often looks, at first glance, like a simple smear: a film on a rock in a stream, a slick on a pipe, a faint fuzz on a tooth. Under the microscope it is something else entirely: a crowded city of microbes embedded in a substance they manufacture outside their bodies. That substance is the extracellular matrix, and it is the reason a dose of antibiotic that would easily wipe out free-swimming bacteria can leave a mature biofilm bruised but alive.

Calling the matrix "slime" is like calling a beehive "wax." It is a material with architecture, chemistry, and purpose. Some of it is made of long sugars (polysaccharides) that hold water like a gel. Some is protein: structural fibers, sticky adhesins, enzymes. Some is lipid. A striking fraction, in many biofilms, is extracellular DNA (eDNA), released when some cells rupture or actively export it. In human infections, host DNA can be added to the mix as immune cells die at the battle line. The result is not a uniform coating but a meshwork with pores, charged surfaces, pockets, and channels through which water and nutrients can move.

Antibiotics enter this environment the way smoke enters a dense forest: they do not meet a smooth open space, but a landscape of surfaces that can snag, absorb, and transform them. For some drugs, the matrix is a modest delay. For others, it behaves like a sponge with a chemical mind of its own, soaking up molecules at the surface and leaving the deeper residents breathing easier.

The simplest protection is purely physical. An antibiotic molecule in fluid reaches a biofilm and begins to diffuse inward, driven by the random jostling of thermal motion. Diffusion is not magic; it depends on pathways being open. The matrix narrows those pathways. It increases viscosity. It turns a straight shot into a zig-zag. If the antibiotic is large, or if the pores are small, diffusion slows. If cells are densely packed and the matrix is thick, the drug may be spent or altered before it ever reaches the bottom.

Yet physical hindrance is only half the story, and often not the most important half. The more subtle defense is chemical: the matrix can bind antibiotics, concentrate them in particular zones, or even dismantle them. The biofilm, in effect, does not merely *block* the weapon; it *handles* it.

Charge is one of the key handles. Many antibiotics carry electrical charges, and so do many matrix components. Aminoglycosides such as tobramycin and gentamicin are positively charged under physiological conditions. A great deal of biofilm matrix material is negatively charged: eDNA, certain polysaccharides, and acidic proteins. Opposites attract. When a cationic antibiotic meets an anionic matrix, it can be pulled out of the flowing water and held in place like static-clung dust.

This is not an abstract chemical curiosity; it shapes the geography of treatment. Experiments with *Pseudomonas aeruginosa*—a bacterium notorious in chronic lung infections and in some wound infections—have shown that tobramycin can accumulate at the periphery of a biofilm rather than distributing evenly through it. In practical terms, the outer layers receive a harsh dose while deeper cells experience a much weaker exposure, sometimes weak enough to survive. Meanwhile, a more neutrally charged drug, such as ciprofloxacin, may penetrate more readily. The "fortress" is not built of a single material that stops everything; it is built of selective filters that treat each incoming molecule differently.

One can picture this as a kind of chemical customs checkpoint. A neutral traveler passes with relatively few questions. A highly charged traveler is stopped, searched, delayed, and sometimes detained indefinitely.

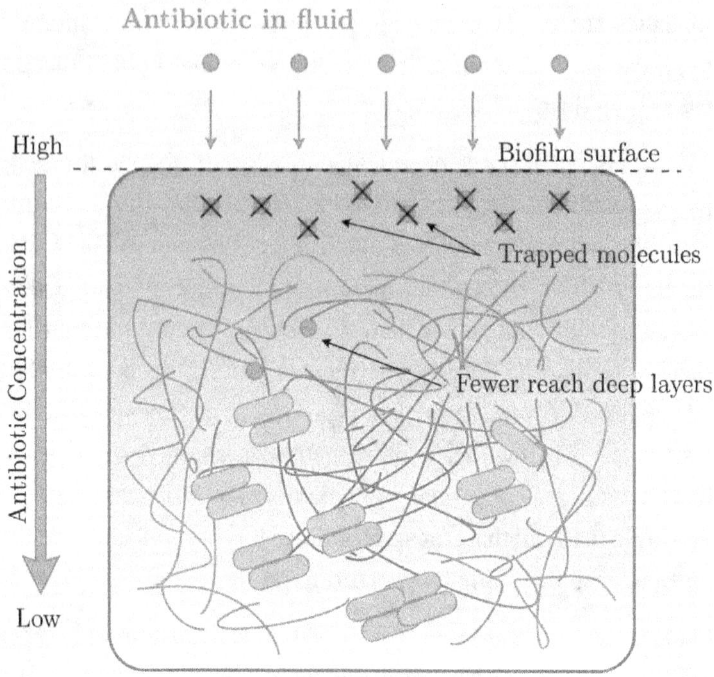


Cells embedded in matrix

Binding does not have to be purely electrostatic. Some antibiotics are hydrophobic; they prefer oily environments to water. Many matrix components have hydrophobic patches. Proteins can present binding sites. Polysaccharides can create microdomains with distinct local chemistry. The matrix is, from the drug's point of view, a maze of surfaces offering opportunities to stick, stall, and accumulate. Even if every molecule *could* in principle diffuse inward, the rate can become so slow that the drug is effectively neutralized by time.

Time matters because antibiotics are not always stable. Some are chemically fragile; some are actively destroyed by enzymes. Biofilms can take advantage of both. A classic example is the family of enzymes known as beta-lactamases, which cleave the beta-lactam ring found in penicillins and related drugs. Bacteria can secrete these enzymes into the surrounding matrix. If enough enzyme accumulates near the surface, incoming antibiotic molecules can be dismantled before they reach inner cells. This creates a situation that feels unfair from the clinician's viewpoint: the drug is present in the patient's bloodstream or in the fluid bathing a wound, but the concentration that matters—the concentration at the target cell—is far lower.

One can see why early researchers who grew bacteria in well-mixed laboratory flasks were repeatedly surprised by medical reality. A culture of free-living bacteria behaves like a crowd in an open field: everyone is equally exposed to the same dose. A biofilm behaves like a city with thick walls and twisting streets: exposure depends on where you live.

A concrete historical thread runs through infections on medical devices. In the late twentieth century, clinicians increasingly recognized that certain infections seemed to cling to catheters, prosthetic joints, and heart valves with stubborn persistence. It was not simply that bacteria were present; it was that they were present *as a community attached to a surface*. J. William Costerton and colleagues helped crystallize this idea by arguing that many chronic infections involved biofilms rather than solitary cells. The observation was not merely descriptive. It explained a repeated clinical pattern: antibiotics might reduce fever and lower bacterial counts in fluid samples, yet the infection returned as soon as treatment stopped, because the true reservoir sat protected on a surface, wrapped in matrix.

The matrix's chemical tricks are especially vivid in *Pseudomonas aeruginosa* infections in cystic fibrosis. Over years of colonization, this bacterium can shift into a "mucoid" form that produces large amounts of alginate, a polysaccharide that gives colonies a slimy appearance. Patients and clinicians have long wrestled with the paradox that inhaled antibiotics can be delivered directly to the lungs in high doses and still fail to fully clear the infection. Alginate is not the whole explanation, but it illustrates the logic: a thick polysaccharide gel can slow diffusion and create a binding environment for certain positively charged drugs. In laboratory studies from the 1990s, purified alginate was shown to impede the movement or activity of aminoglycosides, and enzymes such as alginate lyase could partially undo that barrier, improving diffusion. The details vary by strain and by drug, but the lesson is durable: the matrix can be an active participant in pharmacology, not a passive backdrop.

Extracellular DNA adds a further twist, because it does more than bind molecules; it can reshape the local ionic environment. DNA is densely negatively charged. In a biofilm, eDNA can chelate, or grab, positively charged ions such as magnesium and calcium. Those ions are not trivial decorations. They help stabilize bacterial outer membranes and influence regulatory systems that bacteria use to sense stress. When eDNA mops up these cations, it can create a kind of local mineral famine. Bacteria interpret that famine as danger and may respond by remodeling their cell surfaces in ways that reduce the effectiveness of cationic antimicrobials. The matrix, in this sense, does not merely intercept the drug; it changes the physiology of the inhabitants in a direction that makes them harder to kill.

This is the point where the fortress metaphor can mislead if taken too literally. A stone wall is inert. The biofilm matrix is closer to a living material, continually built and remodeled, infused with enzymes, threaded with DNA, and capable of altering the chemistry around it. It is both a barrier and a reaction vessel.

The "sponge" image is useful because it highlights selectivity. A kitchen sponge does not block all liquids equally; it holds onto some substances and lets others pass, depending on how they interact with its pores and surfaces. Similarly, the matrix does not create a universal blockade against every antibiotic. Some drugs slip through relatively quickly. Others get delayed. Some are inactivated en route. The outcome depends on the match between a drug's properties (charge, size, hydrophobicity, chemical stability) and a particular biofilm's matrix composition and density.

That variability has practical consequences. It helps explain why one antibiotic can work well against a biofilm infection in a particular context while another, equally potent in standard laboratory testing, fails. It also helps explain why combinations sometimes outperform single agents: one drug may penetrate well while another is trapped; one may damage the outer layers, changing permeability for the next. It suggests why clinicians sometimes use very high local doses (for example, antibiotic-impregnated materials) to overwhelm binding sites, much as a saturated sponge eventually drips. And it suggests why some emerging strategies aim not only to kill bacteria but to disrupt the matrix itself: breaking the mesh, cutting the DNA, or dismantling key polysaccharides so that antibiotics can reach their targets at meaningful concentrations.

There is also a darker irony in the matrix's relationship with the immune system. When immune cells attack a biofilm, they can rupture and spill their contents, including DNA, into the surrounding environment. That DNA can become raw material for the biofilm matrix. The defenders can be built, in part, from the debris of the battle.

If the matrix were only a physical wall, one might imagine simply increasing the dose until the drug pushes through. The chemical sponge makes this far more complicated. Increase the dose of a positively charged antibiotic, and the outer matrix may bind more of it, soaking it up like a thirsty cloth. Increase the dose enough, and perhaps the binding sites saturate and penetration improves—but the required concentration may be too toxic for the patient. Meanwhile, enzymes embedded in the matrix can quietly dismantle incoming molecules, and local chemical conditions can shift in ways that change how drugs behave. A biofilm is not merely hard to hit; it changes what it means to be hit.

Underneath these interactions is a simple, unsettling idea: antibiotics were largely developed and tested against bacteria behaving as isolated individuals in well-mixed fluids. Biofilms force drugs into a different world, one with geometry and materials science, with adsorption and catalysis, with microenvironments and gradients. In that world, the first bacteria a drug meets are not necessarily the most important ones. The outer layers can act as sacrificial tissue, absorbing the blow, binding the molecules, and buying time. Deep inside, cells can remain effectively untouched, not because they have special genes, but because the matrix has quietly rewritten the rules of access.

## 6.2 Sleeping Through the Storm

An antibiotic can do everything right and still fail.

It can arrive at the infected tissue in a concentration that looks heroic on paper. It can seep into the biofilm's outer layers. It can kill thousands, sometimes millions, of bacteria. The patient's fever drops; the swelling eases; laboratory tests improve. Then treatment stops, and the infection quietly rebuilds itself as if the drug had never visited.

One explanation is simple geography: some cells were never reached. Another is more unsettling: some cells were reached, but the drug could not *use* them. Many antibiotics are tools designed for busy machines. If the machines are switched off, the tools have nothing to grip.

The term *persister cell* refers to a bacterium that survives antibiotic exposure not because it carries a special resistance gene, but because it has entered a physiological state in which the drug's target is idle. "Physiological" here means something like mood or metabolism rather than identity. The persister is not a different species, not even necessarily a mutant. It is the same organism running at a different speed, like a factory on a night shift with the lights down.

This is a difficult idea to accept because it cuts against a familiar moral narrative about antibiotics. We like to imagine a direct contest: a drug is a weapon, a bacterium is an enemy, and survival implies armor. Persisters survive by doing something less cinematic. They wait.

Antibiotics differ in what they attack, but many of the most widely used classes are at their best against cells that are actively growing and dividing. Beta-lactams (penicillins and their relatives) interfere with cell-wall construction, a process that is frantic during division. Fluoroquinolones disrupt DNA replication. Aminoglycosides jam protein synthesis. If a bacterium is not building a wall, not copying DNA, not making proteins at a rapid clip, then the drug's carefully designed sabotage can become blunt.

Dormancy does not mean invincibility. A sleeping bacterium still needs to maintain a minimum level of order: membranes must hold together; ions must be managed; essential molecules must not decay beyond repair. Some antibiotics and host defenses can exploit those necessities. But many drugs used in everyday practice are, in effect, optimized for the busy hours of microbial life.

How does a biofilm create sleepers?

Partly by design, and partly as a consequence of living in a crowd. A biofilm is not a well-stirred soup; it is a layered structure where resources arrive from the outside. Oxygen is a strong example because it is both vital and quickly consumed. In a mature biofilm, oxygen can be depleted within a surprisingly short distance from the surface. Microelectrode measurements in *Pseudomonas aeruginosa* biofilms in the early 2000s made this visible in a way that is still startling: a thin band near the surface can be oxygen-rich and metabolically active, while deeper regions become oxygen-poor or fully anoxic. In those deeper layers, cells shift toward slow growth, fermentation-like metabolism, or near-standstill. Active protein synthesis, in some experiments, is largely confined to a narrow zone near the edge where oxygen and nutrients are most available.

The consequences are not abstract. Imagine an antibiotic designed to stop DNA replication. Near the surface, where cells are dividing, it can be devastating. Deeper down, where cells are barely replicating at all, the same antibiotic may find few moving targets. The biofilm becomes a community in which different neighborhoods experience different pharmacology.

It is tempting to think of persisters as a rare elite, a tiny squad trained to endure. In reality, persister formation can be both rare and common, depending on conditions. In a rapidly growing planktonic culture, persisters may be a small fraction. In the stressed, nutrient-limited, oxygen-starved interior of a biofilm, the fraction can rise dramatically, not because a genetic switch has spread through the population but because the environment nudges many cells toward low activity.

A concrete clinical example makes the stakes plain: chronic infections in cystic fibrosis. The lungs of a person with cystic fibrosis can become colonized by *Pseudomonas aeruginosa* in communities that behave like biofilms, embedded in thick mucus and host material. Antibiotics can relieve acute exacerbations, sometimes dramatically. Yet eradication is notoriously difficult once chronic infection is established. The bacteria are not always genetically resistant to every drug used; rather, the living conditions in the lung can repeatedly generate pockets of low metabolism where killing is inefficient. The patient is left with a cycle of improvement and relapse, a pattern that feels less like a clean battle and more like trying to stamp out embers in wet wood: flames go down, but heat remains in protected crevices.

The concept of persisters has a history that reads like a detective story. In the 1940s, when penicillin was still a relatively new wonder, Joseph Bigger observed something odd in *Staphylococcus* cultures: even when penicillin should have killed the bacteria, a small fraction survived. These survivors were not mutants in the usual sense; when regrown, their descendants were mostly sensitive again. Bigger's term "persisters" captured the mystery: something about a subset of cells allowed them to persist through treatment without inheriting a permanent advantage.

For decades, persisters were a scientific curiosity, easy to dismiss as laboratory noise. The rise of chronic, device-associated, and biofilm-related infections changed that. Once clinicians faced infections that reliably returned after "successful" antibiotic courses, the persister idea became a missing piece of the puzzle. Kim Lewis and others, writing in the 2000s and 2010s, argued persisters were central to relapse and chronicity. The appeal of the explanation was its simplicity: antibiotics do not fail because bacteria are stronger; they fail because bacteria are *inactive*.

Simplicity, however, is not the same as completeness. Persisters are real, but they are not a single, easily targeted entity. Multiple routes can lead a bacterium into dormancy: stress responses, toxin-antitoxin systems (pairs of genes where one product can inhibit growth), stringent responses to starvation, energy depletion, oxidative stress, and more. Different bacteria may rely on different mechanisms. Even within one species, different conditions may generate persisters by different paths. The term describes an outcome—survival through temporary inactivity—more than it describes a single mechanism.

That diversity is frustrating for drug developers, because it makes it hard to hunt for one "persister gene" and design a bullet for it. A biofilm does not need a single lock when it can build doors with many latches.

The physical realities of a biofilm further complicate the picture. In the interior, not only oxygen but also pH and waste products can vary. Acidic pockets can change the charge state of antibiotics and reduce uptake. Limited nutrients can lower ATP (the cell's energy currency), and some antibiotics, notably aminoglycosides, depend on energized membranes for efficient entry. Low energy can therefore make cells less permeable to the very drug meant to kill them. The biofilm's chemistry, architecture, and physiology form a feedback loop: the environment slows metabolism; slow metabolism alters drug uptake and target activity; reduced killing allows the community to maintain the conditions that keep metabolism slow.

The idea of "sleeping through the storm" risks sounding like a deliberate strategy, as if bacteria decide to play dead. There is no need to imagine intention. In a large population under stress, a spread of metabolic states is inevitable. Some cells will be actively growing, some will be struggling, some will be almost inert. Antibiotics then act as a harsh filter that preferentially kills the active fraction and leaves the inert fraction behind. After the drug pressure lifts, the survivors repopulate. What looks like planning is often the mathematics of survival in a heterogeneous crowd.

That heterogeneity has an emotional edge when it plays out in human bodies. Consider a person with a prosthetic joint infection. Treatment can involve weeks of antibiotics and sometimes surgery, yet infections can recur because bacteria on the implant live in a biofilm. Within that film, there may be cells that have not divided in days, tucked into a low-oxygen niche behind a mound of matrix and debris. They are not numerous, but they are sufficient. When antibiotics stop, they wake, divide, and seed the surface again. A patient experiences this as betrayal: a long course of medication, side effects, disruption of life, and then the same pain and swelling return. The microbial explanation is indifferent, but understanding it can shift the clinical logic from "stronger drugs" to "different strategies."

One strategy is to use antibiotics that can kill non-growing cells more effectively, though options are limited and depend on the organism. Another is to use combinations: one drug might reduce the actively growing population while another targets a different vulnerability, or an adjuvant might alter the environment enough to rouse sleepers into a vulnerable state. A more radical approach is to attack the biofilm's habitat: disrupt the matrix, change local oxygenation, or physically remove colonized surfaces. In many device-related infections, removing the device remains one of the most reliable ways to end the cycle, a blunt but telling fact about how well protected these microbial neighborhoods can be.

Persisters also reshape how one thinks about the word "susceptible." In the clinic, bacteria are often classified by whether their growth is inhibited by a particular concentration of antibiotic in a laboratory test. That test is usually performed on actively growing planktonic cells. A persister can be "susceptible" in that formal sense and still survive treatment because susceptibility is measured against growth, and persisters are defined by the refusal to grow. The label can be technically correct and clinically misleading.

There is an irony here that would be comic if it did not cause so much suffering. Modern medicine prides itself on precision: identify the pathogen, measure the minimum inhibitory concentration, choose the antibiotic, calculate the dose. Persisters reveal a loophole. They do not argue with the drug's chemistry; they sidestep it by changing the biological context in which the chemistry matters.

Even the term "dormant" can mislead. A persister is not necessarily in a deep hibernation like a bear in winter. Some persisters may be slowed rather than stopped, maintaining enough metabolism to repair damage and control internal chaos. Others may be in a more profound shutdown. The key point is that their metabolic activity is low enough that the drug's killing mechanisms are blunted. That is why persisters can survive doses that kill their neighbors, and why, once regrown, they usually produce offspring that are just as sensitive as the original population. The difference was never in the blueprint. It was in the state.

The most haunting aspect of persisters is their quietness. Genetic resistance announces itself by growth in the presence of antibiotic, a clear and alarming signal. Persisters leave no such banner. During treatment, bacterial numbers may fall. Cultures may even become negative. Symptoms may ease. The surviving cells are few and often hard to detect, like a small group sheltering in a basement during a siege. When they re-emerge, it can look as if a new infection has arrived, when in fact it is the old one resuming business.

A biofilm's defense, then, is not only the matrix's ability to slow and bind incoming drugs, but also the ability of the community to generate physiological diversity, including pockets of near-sleep. In that sense, the biofilm is less like a uniform army and more like a society with varied occupations and schedules. Some citizens are out in the street when the storm hits. Others are indoors, lights off, waiting for the wind to die down.

## 6.3 Tolerance, Not Resistance

A curious misunderstanding sits at the heart of many arguments about "superbugs". When an infection refuses to clear, the instinct is to blame *resistance*: bacteria have evolved, the drug no longer works, the microbe has outsmarted medicine. Sometimes that is exactly what has happened. Yet in many stubborn, relapsing infections associated with biofilms, the more accurate word is *tolerance*. The difference is not academic. It changes what tests can tell you, what a successful treatment looks like, and why a patient can feel better on antibiotics only to fall ill again once they stop.

Resistance is genetic. A resistant bacterium carries heritable changes—mutations or acquired genes—that let it grow even while the antibiotic is present. If you take that bacterium, culture it in a laboratory, and test it tomorrow or next week, it will still be resistant. The classic yardstick here is the MIC, the minimum inhibitory concentration: the lowest concentration of a drug that prevents visible growth in a standardized test. Resistance typically raises the MIC. The bacteria do not merely survive; they can *multiply* at concentrations that used to hold them back.

Tolerance is different. A tolerant bacterium may be just as genetically susceptible as its neighbors. It may not grow in the presence of antibiotic. It may even be damaged by the drug. But it survives long enough that, when the antibiotic is removed, it can recover and repopulate. Tolerance is often reversible and dependent on circumstances: nutrient availability, oxygen levels, stress responses, and the architecture of the microbial community. In biofilms, those circumstances are not occasional accidents; they are built into the environment.

One way to feel the distinction is to imagine a snowstorm and a roadblock. Resistance is a car with snow tires and a powerful engine driving straight through the storm. Tolerance is a car parked in a garage until the storm passes. The outcome looks similar from a distance—the trip continues later—but the mechanism is entirely different. The resistant cell is actively growing under antibiotic pressure; the tolerant cell is not growing, merely surviving.

Clinically, this can be maddening. A physician may prescribe an antibiotic based on a laboratory report that states the organism is "susceptible". The patient improves, and the course ends. Weeks later, the infection returns. From the patient's point of view, the drug "didn't work". From the laboratory's point of view, the drug "should work". Both perspectives can be correct if the organism is genetically susceptible but living, during treatment, in a physiological state that makes killing inefficient.

The early history of antibiotics encouraged the resistance-centered worldview. Penicillin's dramatic success in the 1940s created a powerful expectation: give the right drug at the right dose, and the infection collapses. When resistance did arise—and it did, quickly—it offered a satisfying explanation. Alexander Fleming himself warned, presciently, that underdosing could select for resistance. Later decades provided textbook examples: the spread of penicillinase-producing *Staphylococcus aureus*, the evolution of methicillin-resistant *S. aureus* (MRSA), the rise of extended-spectrum beta-lactamases and carbapenemases among Gram-negative bacteria. These are real genetic arms races, with public health consequences.

Biofilms add a different kind of disappointment. Here, the failure can occur even without a decisive genetic shift. A biofilm can take a population of susceptible bacteria and make it behave *as if* it were resistant, at least from the point of view of a clinician waiting for the infection to clear. The microbe has not necessarily changed its genome. It has changed its way of life.

That change creates a trap in language. The word "resistant" is often used informally to mean "hard to kill". In casual conversation it makes sense: if antibiotics don't clear the infection, the bacteria must be resistant. But in microbiology, resistance is a specific genetic property, and confusing it with tolerance blurs the logic of both diagnosis and treatment.

A practical consequence is that the tests most often used in clinical laboratories were designed to measure *inhibition of growth*, not *eradication of a structured community*. MIC testing takes bacteria, usually grown in broth so they are metabolically active, and exposes them to a range of drug concentrations. The endpoint is simple: which concentration stops visible growth? That is a sensible question when bacteria behave as free-living cells. It is a less complete question when bacteria are embedded in a biofilm, where many cells may not be growing much anyway.

If one wants to ask the more clinically relevant question for a biofilm infection—what does it take to *eradicate* the community, not merely stop it from expanding?—a different measurement is needed. This is where the concept of the MBEC enters: the minimum biofilm eradication concentration. It is, roughly speaking, the concentration required to kill a biofilm to the point that it cannot regrow when transferred to fresh conditions. MBEC values can be dramatically higher than MICs for the same organism and drug, not because the bacteria have suddenly become genetically different, but because killing a biofilm is a different task from inhibiting growth in broth.

Standardization matters here. Researchers can build almost any "biofilm model" they like, which makes comparisons difficult. Efforts such as ASTM E2799 aim to standardize methods for growing biofilms and testing antimicrobials against them, allowing more meaningful comparison across labs. Even with standardization, the general message remains uncomfortable: the number printed on a susceptibility report may not predict what happens in a chronic, biofilm-associated infection.

The distinction between resistance and tolerance also clarifies a clinical puzzle: why infections relapse after apparently adequate treatment. In genetic resistance, relapse may occur because the drug never truly suppressed the population; resistant bacteria kept growing all along, perhaps at a slower pace. In tolerance, relapse can occur even when the drug did suppress growth and killed a large fraction of cells. Survival, not growth, was enough.

A classic setting is infection on a foreign surface: a urinary catheter, a central venous line, a prosthetic joint, an artificial heart valve. These infections are often described as "recalcitrant", a word that sounds almost moral, as if the bacteria are being stubborn. The stubbornness is frequently a property of the biofilm lifestyle. Remove the device and the infection may resolve with antibiotics that previously seemed ineffective. The genome in the lab dish may not have changed; the habitat in the body did.

There is a human story embedded in that logic. Modern medicine has filled the body with helpful materials: plastics and metals that keep patients alive, restore movement, deliver drugs, and repair damaged anatomy. Those materials provide new real estate for microbes. A catheter is a smooth, reliable surface in a warm, nutrient-rich environment with a steady flow of fluid. A prosthetic joint is a stable platform that cannot shed its surface the way skin does. Biofilms treat these innovations as invitations. The resulting infections can be devastating not only because they are hard to treat, but because the treatment can involve removing or replacing the very device that was meant to improve life.

The resistance-tolerance distinction also helps make sense of why "more antibiotic" is not always the answer. With genetic resistance, increasing dose can sometimes overcome low-level resistance, though toxicity limits this strategy. With tolerance, increasing dose may have diminishing returns if the surviving cells are not dying because the relevant targets are inactive, or because the biofilm environment prevents adequate exposure. The intuitive response—turn up the volume—can fail when the problem is not volume but context.

This is a good moment to place the key ideas in a single picture. The difference between resistance and tolerance is easier to grasp as a timeline than as a definition.

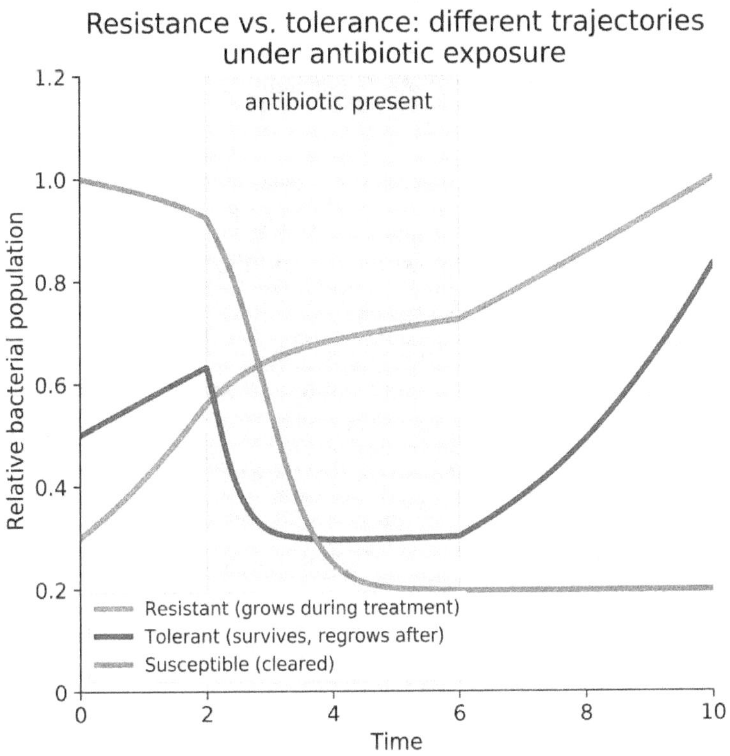


The chart makes a quiet but decisive point. Resistant bacteria remain *biologically ambitious* under drug pressure. Tolerant bacteria remain *biologically cautious.* The drug may suppress them, even injure them, but it does not eliminate them. Once the drug is gone, normal life resumes.

One might ask whether tolerance matters beyond biofilms. It does.

Persister cells can arise in planktonic cultures too, and tolerance can contribute to relapse in infections that are not classically biofilm-based. Yet biofilms are an amplifier. Their structured environment reliably produces slow-growing and stressed subpopulations, and their matrix can shield and shape exposure so that "enough survives" becomes a recurring outcome. The combination is powerful: a chemical environment that reduces effective drug concentration at depth, and a physiological environment that reduces the vulnerability of cells even when drug arrives.

There is also an evolutionary twist. Tolerance is not the same as resistance, but it can be a staging ground for resistance. A drug that kills rapidly leaves few survivors and therefore few opportunities for resistant mutants to arise. A drug that fails to eradicate, leaving a lingering population under intermittent pressure, provides time and repeated selection opportunities. Chronic biofilm infections can therefore become arenas where tolerance and resistance interact, sometimes sequentially. The organism may begin as genetically susceptible yet tolerant in a biofilm; over time, under repeated antibiotic courses, resistant mutants can emerge and spread. The end point looks like "resistance" in the everyday sense, but the path often ran through tolerance first.

This is why the language matters. If every treatment failure is framed as resistance, the response tends to be a search for new antibiotics, stronger drugs, broader coverage. That search is vital, but it is not sufficient. Tolerance suggests additional strategies: disrupting the biofilm habitat, timing and dosing regimens that account for dormancy, combining antibiotics with agents that break down matrix components, designing diagnostics that measure killing rather than growth inhibition, and making clinical decisions (such as device removal) that recognize where the microbial reservoir actually lives.

A final, concrete example returns to the experience of patients with chronic wounds. In diabetic foot ulcers, bacteria can form biofilms on the wound bed. Antibiotics may reduce signs of infection, yet healing stalls. Debridement—the physical removal of dead tissue and biofilm-laden material—can sometimes succeed where antibiotics alone do not. It is an old surgical practice with a modern microbiological explanation: you are not simply removing "dirt", you are removing a structured community whose tolerance mechanisms depend on its physical and physiological environment. The action is blunt, but its logic is precise.

Tolerance is easy to overlook because it hides behind the familiar apparatus of susceptibility testing and the language of resistance. But once noticed, it reframes the question. The problem is not always that the drug is too weak. Sometimes the problem is that the target has chosen, for a while, not to be a target at all.

## 6.4 Frustrating the Phagocytes

If antibiotics are the tools of modern medicine, the immune system is its older, more adaptable craft. It does not need a prescription, it does not wait for a laboratory report, and it does not run out when the pharmacy closes. It recognizes patterns of danger, floods tissues with defenders, and, when "all goes well," clears an infection with a mixture of force and finesse.

Biofilms, unfortunately, are built to make "all goes well" less likely.

The immune system has many weapons, but the most iconic are the *phagocytes*—cells whose job is to find microbes and eat them. "Phagocyte" comes from Greek roots meaning "devouring cell," and the image is accurate: a neutrophil or macrophage can wrap itself around a bacterium, pull it inside, and destroy it in a chemically harsh compartment. This process, phagocytosis, is one of the body's cleanest solutions to infection. It removes invaders rather than merely injuring them, and it does so with a kind of physical certainty. A bacterium inside a phagocyte has few escape routes.

Biofilms exploit a limitation that is almost embarrassingly simple. A phagocyte can eat a bacterium. It can sometimes eat a small clump. It cannot eat a city.

A mature biofilm is often larger than a single immune cell and firmly attached to a surface. When a neutrophil encounters such a structure, it may attempt to engulf it anyway, spreading its membrane around the mass. The attempt fails because the target is too big or too anchored. The immune cell becomes stuck in a half-swallow, committed to an action it cannot complete. This is called *frustrated phagocytosis*, and it is one of the crueler tricks in the biofilm repertoire. The phagocyte responds as if it has captured its prey: it releases reactive oxygen species, proteases, and other toxic molecules meant for the inside of an engulfing compartment. But because the compartment never seals, those toxins spill into the surrounding tissue.

The result is not a heroic last stand of the immune system; it is collateral damage. The biofilm remains largely intact while the host's own enzymes and oxidants chew at nearby cells, connective tissue, and blood vessels. In chronic biofilm infections, this mismatch between effort and effect can drive ongoing inflammation: a persistent, low-grade war that harms the body more reliably than it harms the microbes.

Neutrophils are often the first responders. They are abundant, fast, and disposable in a way that macrophages are not. A neutrophil can die as part of its function, and the body will replace it. That very disposability becomes an opportunity for biofilms. In encounters with *Pseudomonas aeruginosa* biofilms, neutrophils can become immobilized, rounded up, and less able to crawl effectively through the matrix-rich environment. They arrive, they sense danger, they activate—and then they fail to remove the target. They can spend their lethal chemistry on the wrong substrate: the patient's own tissue.

Macrophages, which are longer-lived and more versatile, face similar constraints. They can engulf bacteria and debris; they can coordinate other immune responses; they can help repair tissue. Yet against an adherent biofilm they can also be reduced to ineffective contact, releasing inflammatory signals and enzymes without achieving clearance. The immune system is not absent in biofilm infections. Often it is intensely present, sometimes destructively so.

A concrete example sits in the everyday ache of periodontal disease. The mouth is one of the richest microbial habitats in the body, and dental plaque is a biofilm with remarkable sophistication. Most people live with plaque without catastrophe because the immune system and oral hygiene keep it within bounds. When plaque persists at the gum line, the immune system responds. Neutrophils migrate into the crevice between tooth and gum, and inflammatory molecules accumulate. Over time, that chronic inflammation can damage the supporting structures of the teeth. Bone is resorbed; gums recede; pockets deepen. The biofilm gains more sheltered space, and the immune system's attempt to control it ends up remodeling the mouth in ways that benefit the microbes. This is not merely a problem of "bacteria attacking." It is a problem of an immune response that cannot neatly remove an attached community and instead grinds down the landscape around it.

The story repeats, with higher stakes, in chronic lung infections. In cystic fibrosis, thick mucus and impaired clearance create conditions in which biofilms can persist. The lungs can become filled with neutrophils responding to infection. Those neutrophils release enzymes such as elastase and other proteases, along with reactive oxygen species. These molecules are meant to kill bacteria, but in excess they damage lung tissue, thicken secretions, and worsen lung function. The infection persists; inflammation persists; the tissue becomes a casualty of its own defense.

A biofilm's matrix contributes to this in two ways. First, it acts as a scaffold, holding the microbial community together and physically resisting engulfment. Second, it shapes the chemistry of the battleground. The matrix can slow the diffusion not only of antibiotics but of immune factors: antibodies, complement proteins, antimicrobial peptides. It can bind some of them. It can create local microenvironments that blunt their effectiveness, such as zones of altered pH or low oxygen where certain oxidative killing mechanisms are less efficient.

The immune system also communicates, and biofilms are skilled at jamming signals. Antibodies can coat bacteria and mark them for phagocytosis, a process called opsonization. Complement proteins can punch holes in membranes or tag microbes for destruction. In a biofilm, the outer layers can intercept these tags, and the deeper cells may be spared direct contact. Even when bacteria are coated, a phagocyte still faces the mechanical problem of eating something that is anchored and too large.

There is a grim twist here: the immune system's attack can supply building materials to the biofilm.

Extracellular DNA, a key component of many biofilm matrices, can come from bacteria themselves, but it can also come from the host. When neutrophils die, they can release DNA into the environment. In some contexts neutrophils actively extrude DNA in web-like structures called neutrophil extracellular traps, or NETs, designed to ensnare microbes. These traps are ingenious: rather than swallowing, the neutrophil casts a net laced with antimicrobial proteins to hold pathogens in place.

In the presence of a biofilm, that DNA can be repurposed. DNA is sticky, long, and negatively charged. It can integrate into the matrix, increasing its structural integrity and its ability to bind positively charged molecules. Experiments have shown that DNA from lysed immune cells can increase antibiotic tolerance in certain biofilms, particularly against aminoglycosides. In other words, the immune system's attempt to trap bacteria can thicken the very mesh that helps them endure.

This is one of the most psychologically unsettling aspects of chronic biofilm infections: inflammation can become a form of inadvertent collaboration. The body tries harder, and the biofilm becomes tougher.

The phrase "immune evasion" is sometimes used as if microbes turn invisible. Biofilms rarely need invisibility. They need endurance. They create a situation in which immune cells cannot perform their cleanest maneuver—engulfment—and are instead pushed toward messy chemistry. That chemistry is not harmless to bacteria. Outer layers of a biofilm can be injured, and some cells die. But the biofilm can treat those losses as acceptable. A city can lose a few buildings and remain a city.

A historical anecdote helps frame why this realization took time. For much of the twentieth century, immunology and microbiology developed with a bias toward what could be easily cultured and counted. Bacteria were studied as planktonic cells; immune cells were studied in simplified assays. The idea that bacteria commonly live in attached, matrix-embedded communities in nature and in the body gained traction slowly. J. William Costerton and colleagues, writing around the turn of the twenty-first century, helped popularize the view that biofilms are not rare quirks but a common microbial lifestyle, especially in chronic infections. Once that lens is adopted, many clinical puzzles become less mysterious: why infections cling to catheters, why wounds refuse to heal, why inflammation persists in the presence of organisms that appear "susceptible" in standard tests. The immune system was not failing in the abstract. It was being asked to solve a problem—removing an adherent community—for which its most elegant tools are poorly suited.

Biofilms can also manipulate what the immune system *decides* to do. Some biofilm-associated bacteria alter the molecules on their surfaces, changing how strongly immune receptors recognize them. Some secrete factors that dampen immune activity or misdirect it. *Pseudomonas aeruginosa*, for instance, produces a variety of secreted products that can injure host cells and modulate immune responses. But it is often the architectural reality that sets the stage: even a fully alert immune system cannot easily consume a structure it cannot wrap around.

The idea of "frustration" in frustrated phagocytosis is almost literal. A neutrophil arriving at a biofilm may sense danger, activate its killing machinery, and still be unable to finish the job. The immune cell's own commitment becomes a liability. It is like deploying a weapon that only functions safely when used in a confined chamber, then firing it in the open air.

This helps explain why chronic biofilm infections are often characterized by a particular emotional tone in clinical practice: weary persistence. They are not always dramatic. They may not trigger the overwhelming sepsis of an acute bloodstream infection. Instead they linger: a low-grade pain around a prosthesis, a foul-smelling wound that improves and then worsens, a cough that never fully resolves, gums that bleed a little each day. The immune system is engaged, sometimes intensely, but it cannot clear the attached reservoir. The patient's tissues bear the cost of ongoing skirmishes.

Understanding this changes the moral of the story. The immune system is not simply too weak, and the bacteria are not simply too strong. The conflict is mismatched. Phagocytosis is a superb solution for individual invaders and small clusters. Biofilms are structured to escape that scale of combat, pushing the immune response toward strategies that are powerful but indiscriminate.

There is a practical lesson in this, one that clinicians have long acted on even before the microbiological explanation was fully clear: sometimes the best way to help the immune system is to change the battlefield. Drain an abscess. Remove a colonized catheter. Debride a chronic wound. Clean a tooth surface. These interventions do not replace immunity; they restore a situation in which immunity can work. When the adherent mass is broken up and reduced, phagocytes can again do what they evolved to do: swallow, contain, and destroy.

The biofilm, meanwhile, continues to take advantage of every failed engulfment. Each frustrated phagocyte leaves behind a smear of enzymes, oxidants, and often DNA. Those residues can inflame tissue, feed the matrix, and deepen the shelter. In chronic infection, the tragedy is not only that microbes persist, but that the body's attempt to expel them can slowly make the environment more hospitable to persistence: damaged surfaces, exposed binding sites, thickened secretions, and a steady supply of molecular debris that, in the wrong hands, becomes mortar.

Chapter 7

# The Clinical Battlefield

*Modern medicine excels at killing free-floating bacteria, yet it often finds itself powerless against the entrenched fortresses of the biofilm. This chapter moves from the sterile operating theater to the scarred landscape of the chronic wound, revealing how microbes hijack our own medical devices and immune responses to build persistent, drug-resistant settlements.*

## 7.1 The Race for the Surface

A modern hospital contains a quiet miracle: parts of the human body are replaced, rerouted, or reinforced with objects made in factories. A hip becomes titanium and polyethylene. A heartbeat is steadied by a pacemaker. A kidney failure is bridged by a catheter that can sit in a vein for weeks, delivering drugs precisely where they are needed. These devices work because the body is astonishingly tolerant of foreign materials-up to a point.

The problem is that the body is not the only inhabitant invited to this new surface.

The moment an implant meets living tissue, it enters a brief, decisive contest that J. W. Gristina named in the late 1980s: the *race for the surface*. The phrase sounds almost playful until you realize what is being raced for. Not pride, not performance-territory. A few square centimetres of polymer or metal that did not exist in the body yesterday becomes, today, the most valuable real estate in the local ecosystem. Whoever claims it first shapes what follows: uncomplicated healing and long-term function, or a protected bacterial stronghold that can flare into recurrent infection months or even years later.

It is tempting to imagine that infection begins when bacteria arrive, and ends when antibiotics kill them. Biofilms complicate that story. The critical step is often earlier and simpler: *attachment*. If microbes attach before the host has effectively sealed the surface into its own tissue, they can build a community that is difficult for immune cells to reach and difficult for drugs to clear. The device then stops being a passive object and becomes an ecological platform-a hard, clean shoreline on which microbes can build a defended city.

What makes the timing so cruel is that neither side needs to win by much. A tiny head start can decide the future.

Consider what the body does to a new device in the first minutes. Blood and interstitial fluid bathe the surface, coating it with proteins almost immediately. Engineers call this "conditioning film," but "film" can mislead: it is not a uniform varnish. It is a shifting layer of albumin, fibrinogen, fibronectin, and other molecules that adsorb, detach, unfold, and rearrange depending on the chemistry and roughness of the material. The surface that the surgeon implants is not the surface that cells and microbes actually touch. They meet a protein-draped landscape whose features are measured in nanometres.

Host cells respond to that landscape with a familiar purpose: close the wound, restore barriers, and incorporate the foreign object into stable tissue. On a hip implant, bone-forming cells and their supporting cast want intimate contact with the metal so that the joint becomes mechanically integrated. Around many devices, connective tissue grows to wall off and stabilize what cannot be digested. Either way, the body's strategy is to *occupy the interface.* When host cells and matrix proteins cover the device, it becomes harder for microbes to find bare footholds and harder for them to remain if they do.

Microbes, meanwhile, are exquisitely prepared for this exact moment. The bacteria most notorious in device infections are not necessarily the most violent; they are the most opportunistic. *Staphylococcus epidermidis*, for example, is a common resident of healthy skin. In its usual habitat it is often harmless, even helpful. But it is also a specialist in clinging to synthetic surfaces and building biofilms. A catheter can be contaminated by organisms already living on the patient's skin; a prosthetic joint can be seeded during surgery by a few cells that never belonged in the wound. The numbers can be small. The consequences can be large.

Attachment is not a single act, like a suction cup. It unfolds in stages. A bacterium drifting near a surface may first "dock" loosely using weak physical forces: electrostatic attraction, hydrophobic effects, van der Waals interactions. At this stage it can still be knocked away by fluid flow or by a well-timed immune response. If it remains, it begins to "lock" itself down by expressing adhesins-surface proteins that bind to host proteins in the conditioning film, such as fibrinogen or fibronectin. In other words, the microbe does not need to recognize titanium. It recognizes the body's own molecules that have coated the titanium.

Then comes construction.

Once a few bacteria settle, they can begin producing extracellular polymeric substances, often shortened to EPS: a mesh of polysaccharides, proteins, lipids, and extracellular DNA that forms the biofilm matrix. The matrix is sometimes described as "slime," and in a literal sense it is slimy-hydrated, sticky, deformable. But "slime" makes it sound accidental, like something spilled. The matrix is built. It is scaffolding and armour at once. It holds cells close enough to cooperate, traps nutrients, binds ions, and creates microenvironments where oxygen, acidity, and drug concentrations differ from the surrounding tissue. It also changes the physics of the battlefield: immune cells that might grab and engulf a single bacterium now confront a cohesive, elastic mass that resists removal.

The race for the surface is therefore not simply host cells versus bacteria in a head-to-head duel. It is host tissue attempting to seal and domesticate a foreign object, while microbes attempt to establish a protected community before that seal forms.

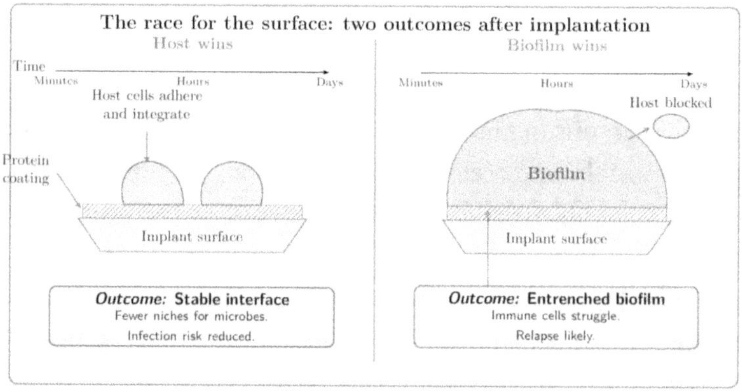


The picture is simplified, but the logic is accurate: the contest happens early, and it is often decided by events too small to notice. A few bacteria that slip past sterile technique. A brief lapse in catheter handling. A micro-scratch on plastic that provides shelter from shear forces. A thin sheath of fibrin that forms as the body responds to injury-useful for healing, but also a sticky welcome mat for microbes that know how to grip it.

Clinical history is full of episodes where medicine learned, painfully, that a device can be a cure and a risk in the same breath. One striking example is the rise of long-term intravascular catheters. These lines transformed cancer care, intensive care, and dialysis. They also created a new category of infection in which the "site" of disease was not a tissue in the usual sense but a surface that bridged skin and bloodstream. Bacteria could migrate along the catheter from the skin entry point, or be introduced at the hub during handling, or seed the device from a transient bloodstream infection. Once attached, they could persist in a biofilm that shed cells intermittently into the blood-turning the catheter into a reservoir rather than a one-time exposure. The clinical pattern was peculiar and maddening: fevers that returned, blood cultures that cleared and then turned positive again, symptoms that eased on antibiotics but reappeared when the course ended. It was not that the drugs never worked; it was that the surface kept *restarting* the infection.

This is where the race for the surface becomes more than a metaphor. It becomes a practical lens for prevention.

If early attachment is decisive, then the peri-operative period-the hours surrounding implantation-carries disproportionate weight. That is why surgical antibiotic prophylaxis is timed carefully: not "whenever," but when tissue concentrations will be highest at the moment bacteria are most likely to arrive. It is why operating rooms have elaborate sterility protocols, and why a small break in technique can matter even when the incision looks clean afterward. It is why catheter bundles emphasize antisepsis at insertion and meticulous handling afterward. All of these practices are attempts to slow the microbial runner at the starting line.

Yet speed is not the only variable; the track itself matters. A surface is not merely a stage. It is a participant.

Materials differ in hydrophobicity, charge, and micro-topography, all of which influence which proteins adsorb, how they fold, and how accessible they are to binding. Even the same material can behave differently depending on how it is processed: polished, roughened, coated, or aged. Some implant designs intentionally roughen surfaces to encourage bone integration. That roughness can be beneficial for the host, but it can also provide protected niches where bacteria are harder to dislodge. The point is not that roughness is "bad," but that the surface features that help one side may help the other, and the balance can be context-dependent.

Microbes also change their behavior when they sense a surface. Many bacteria switch on gene programs that make attachment more stable and communal living more likely. Communication molecules-chemical "words" passed between cells-allow a sparse population to behave as if it were dense, coordinating matrix production, nutrient acquisition, and stress responses. The biofilm is therefore not just bacteria stuck to a surface; it is bacteria *responding* to the fact that they are stuck, turning a precarious foothold into architecture.

Once the matrix is established, the immune system encounters a problem of geometry. Neutrophils and macrophages are designed to engulf targets roughly their own size. A biofilm community can be much larger, and it is physically anchored. Immune cells can press against it, release reactive molecules and enzymes, and still fail to remove it. The assault can harm nearby host tissue while leaving the biofilm largely intact. This is one reason device infections can smoulder: the immune system is not absent, but misapplied, forced to fight a form of microbial life it did not evolve to cleanly eliminate.

Antibiotics face a different kind of frustration. The matrix can slow diffusion, but the more important barrier is physiological. Deep in a biofilm, cells may grow slowly because nutrients and oxygen are limited. Many antibiotics are most lethal to actively dividing cells. Slow growth can therefore look like resistance even when the bacterium has no resistance genes. In addition, a small fraction of cells may adopt dormant states-often called persisters-that tolerate antibiotic exposure and then repopulate when the drug is gone. The result is a familiar clinical paradox: laboratory tests suggest the organism should be susceptible, the patient improves on treatment, and yet the infection returns because the surface community was never truly eradicated.

This is why clinicians speak so often about *source control*. In device-associated infection, the “source” can literally be the device. When a mature biofilm occupies hardware, antibiotics alone may be asked to do the impossible: sterilize a protected, surface-bound community without removing its shelter. In some cases, careful debridement and targeted antibiotics can succeed while retaining an implant, especially if the biofilm is young and the device stable. In other cases, cure requires removal or exchange. For patients, that can mean additional surgery, prolonged rehabilitation, and a new round of risk. For healthcare systems, it means cost and complexity. For microbes, it is simply ecology: a city that cannot be bombed from the outside must be dismantled.

The tragedy is that this entrenched state can begin with an event so small that it feels unfair. A single bacterium on a glove. A few cells from the patient’s own skin. A transient bloodstream seeding from a distant infection. Biofilms turn such small beginnings into lasting consequences because they exploit a property that the body cannot change: an implanted surface does not bleed, does not renew itself, and does not recruit immune cells in the same way living tissue does. It is a hard island with no resident defenders.

The hopeful side of the race for the surface idea is that it suggests leverage. If outcomes are decided early, then early interventions can have outsized effects. This is not limited to antibiotics. It includes surface coatings that reduce protein adsorption or release antimicrobials locally; designs that discourage bacterial anchoring; strategies that accelerate host integration; and clinical practices that reduce bacterial exposure during and after placement. Each approach is an attempt to shift a close contest by a fraction of time or probability.

There is also a humbler lesson: the patient and the microbe do not begin the race on equal footing. The microbe's body plan is built for rapid attachment and communal defense. The human body's plan is built for healing, which is slower by design because it must be careful. A scar that forms too quickly can be as harmful as one that forms too slowly. Bone that remodels haphazardly will fail mechanically. In that carefulness lies vulnerability. Biofilm infections exploit the fact that "good" healing takes time.

When a device remains uninfected for years, it is easy to forget that it once sat at the center of a microscopic contest. When a device becomes infected, it can feel as though misfortune arrived out of nowhere. The reality is more precise and, in a way, more unsettling: a brief window existed when the outcome was being negotiated molecule by molecule. For a moment the implant's surface was undecided territory. Then one side laid down permanent infrastructure, and the rest followed from that first claim.

## 7.2 The Siege Within

An implant infection feels like a crime scene with a single address: the catheter tip, the prosthetic joint, the valve ring. Chronic biofilm disease is harder to localize. The surface is still there, but now it is the body's own tissue: airway mucus, an ulcer bed, a sinus cavity, the folds of inflamed gum. The biofilm is no longer merely *on* the body. It is *in* it, embedded in places where the immune system can arrive quickly yet still fail to finish the job.

That failure is not for lack of effort. In many chronic biofilm conditions, the immune system fights constantly, loudly, and destructively. The tragedy is that the damage often lands on the host.

A useful phrase for this mismatch is *frustrated phagocytosis*. "Phagocytosis" is the process by which immune cells such as neutrophils and macrophages engulf and digest microbes. It is one of the body's most elegant tricks: rather than poisoning everything, a cell wraps itself around a target, pulls it inside, and dismantles it in a controlled chemical furnace. But phagocytosis has an implicit assumption—targets are small enough to be swallowed.

Biofilms violate that assumption. A biofilm can be as cohesive as a wet scab and as stubborn as plaque on a rock. When neutrophils reach a biofilm, they can recognize microbial molecules, swarm, and attempt to eat. They flatten against the matrix, probing for purchase. When engulfment fails, they switch tactics: they release reactive oxygen species, antimicrobial peptides, and enzymes into the extracellular space. These are powerful weapons, designed for short-range use. Outside the controlled interior of a phagocyte, they are indiscriminate. They can chew up collagen, injure epithelial cells, and inflame blood vessels. The biofilm, wrapped in its matrix and often anchored in protective niches, absorbs the assault better than the surrounding tissue. The result looks like a siege in which the defenders keep repairing their walls while the attackers, in frustration, burn the countryside.

Nowhere is this dynamic more famous, and more heartbreaking, than in cystic fibrosis.

Cystic fibrosis (CF) is caused by inherited changes in the CFTR gene, which affects how salt and water move across epithelial surfaces. In the airways, one consequence is that mucus becomes thicker and harder to clear. The lungs have their own cleaning system: tiny beating hairs called cilia move a thin layer of mucus upward, carrying trapped particles and microbes out toward the throat. In CF this escalator slows. Mucus sits, dehydrates, and becomes a difficult terrain where bacteria can settle, sheltered from physical clearance.

The microbial cast in CF has changed over the decades as treatments have improved. Early in life, children often experience infections with organisms such as *Staphylococcus aureus* or *Haemophilus influenzae*. As patients live longer and undergo repeated antibiotic courses, *Pseudomonas aeruginosa* has historically become a dominant, persistent colonizer in many individuals. *P. aeruginosa* is not merely a bacterium that happens to be present; it is a skilled biofilm builder. In the CF lung it can adopt a "mucoid" form, producing a viscous polysaccharide called alginate that thickens its protective matrix and helps it persist.

The essential paradox of CF lung disease is that the lungs are inflamed as if battling an acute infection, yet the infection persists as if the immune system were absent. Sputum from many patients is packed with neutrophils. Their enzymes, meant to dismantle microbes, contribute to airway injury. One of the key enzymes is neutrophil elastase, which can degrade structural proteins and impair the very clearance mechanisms the lung needs. Over time, repeated cycles of inflammation and tissue damage can reshape the airways, widening and scarring them, creating pockets where mucus pools and oxygen gradients form. Those gradients matter: the deeper parts of thick mucus can become relatively low in oxygen, and bacteria can shift into slower metabolic states that are less vulnerable to many antibiotics.

A clinical clue that biofilms are involved comes from the rhythm of symptoms. Patients can experience exacerbations—periods of worsening cough, breathlessness, and fatigue—that improve with antibiotics, only to recur. Antibiotics can reduce bacterial load and dampen inflammation, but they often do not sterilize the airway. The microbial community persists in protected reservoirs and regrows when drug pressure eases. The patient is left not with a single cured-versus-uncured event but with a chronic negotiation: suppress, recover, repeat.

This pattern is not unique to CF. It appears in chronic sinusitis, in some urinary infections, and in chronic wounds. But CF makes the immunological mismatch especially vivid because the lungs are otherwise so well defended. They are constantly exposed to the outside world, yet in healthy people they remain largely sterile below the throat. CF reveals what happens when the physical side of defense—mucus clearance—is impaired: the immune system arrives, but the battlefield geometry favours the biofilm.

A historical shift in how CF infection was understood came with improved imaging and microbiology. For a long time, persistent infection was imagined as a planktonic (free-floating) problem: bacteria in fluid, accessible to antibiotics, perhaps just repeatedly reintroduced. The biofilm perspective reframed persistence as a structural phenomenon. Bacteria were not merely present; they were *settled.* They built an environment that altered oxygen, pH, and nutrient availability, and they occupied niches where the immune system's strongest weapon—engulfment—was poorly suited. This conceptual shift did not magically cure CF, but it changed how clinicians and researchers thought about therapy: the target was not only the microbe's genetics, but its lifestyle.

Biofilm lifestyle also clarifies why chronic infection can persist even when laboratory susceptibility tests suggest antibiotics should work. A standard susceptibility test typically exposes actively growing bacteria in a well-mixed environment to a drug and measures growth inhibition. Biofilm bacteria live differently. Deep within the matrix, growth can be slow; drug penetration can be uneven; local chemistry can inactivate certain antibiotics; and some cells can enter a tolerant state. Kim Lewis's work on persister cells (around 2010) crystallized a concept that clinicians had long suspected: a small fraction of microbes can survive bactericidal antibiotics without being genetically resistant, simply by being temporarily dormant or otherwise physiologically protected. When treatment ends, these survivors can repopulate. "Tolerance" is not a comforting word in infection; it means the enemy can wait out the bombardment.

The immune system can also be tricked. Neutrophils respond strongly to microbial signals, but biofilms can alter those signals, hide them, or present them in ways that keep inflammation simmering rather than resolving. Some bacteria secrete molecules that modulate immune responses; others benefit simply from being clustered and shielded. Even extracellular DNA, a matrix component, has double meaning: it strengthens the biofilm and also comes from dead host cells and neutrophil extracellular traps (NETs), which are webs of DNA and antimicrobial proteins thrown out by neutrophils as a last-ditch attempt to ensnare pathogens. In a biofilm-rich environment, NETs can become more building material than weapon, thickening the battlefield with viscous polymers.

A chronic wound offers a different, equally sobering setting for the same siege logic. Imagine a diabetic foot ulcer. The wound is open to the environment, but it is not merely contaminated; it can become trapped in a prolonged inflammatory state. Reduced blood flow, altered sensation, and impaired immune function in diabetes all contribute, but biofilms add a special kind of stubbornness. Under a glossy surface or within a thin layer of slough, a biofilm can persist as a patchy, microscopic community. Routine swabs and cultures can miss it, because biofilm organisms are not evenly distributed and may not grow readily under standard conditions. Clinicians are then asked to treat what they cannot reliably see.

A widely cited microscopy study from 2008 reported biofilm structures in a majority of chronic wound samples examined, but in only a small fraction of acute wounds. The numbers were striking not because they proved every chronic wound is a biofilm, but because they supported what many clinicians had sensed: chronicity is often associated with an organized microbial presence, not just random contamination. A wound bed can become a habitat where bacteria and fungi cooperate, exchange genes, and share matrix. The immune system can keep the infection from exploding into sepsis while still failing to clear it locally. That containment, paradoxically, is part of the problem: the wound persists, the inflammation persists, and the patient's quality of life erodes.

Here, too, frustrated phagocytosis becomes visible in clinical consequences. Neutrophil enzymes and inflammatory mediators can degrade growth factors and extracellular matrix needed for healing. The wound remains stuck in an early phase of repair, unable to move into orderly tissue rebuilding. The biofilm benefits from the chaos: damaged tissue leaks nutrients, and repeated courses of antibiotics can select for communities that are hardier, not necessarily more resistant in the genetic sense, but more tolerant as structured biofilms.

The logic of management follows the logic of siege warfare, even if clinicians rarely use that phrase. In chronic wounds, debridement—the physical removal of devitalized tissue and microbial burden—can be decisive because it disrupts the biofilm's structure. Unlike a purely chemical strategy, it changes the geometry. It peels away the wall. But biofilms can re-form, sometimes quickly, which is why wound care often combines debridement with strategies that improve blood flow, manage pressure, control glucose, and apply topical agents. The goal is not merely to "kill germs." It is to shift the wound environment away from a stable biofilm habitat and toward one where host tissue can reassert control.

What makes all of this emotionally difficult is that the body is doing what it is supposed to do: send defenders, produce inflammatory signals, mobilize weapons. The patient may even have fevers, elevated inflammatory markers, and pain. Yet the microbes persist, partly because biofilms turn microbial life into a collective phenomenon. A single bacterium is a speck, easily eaten. A biofilm is a neighbourhood with shared infrastructure. It can absorb shocks, shelter vulnerable residents, and regenerate after attacks.

This does not mean chronic biofilm infections are hopeless. The last decade has brought a revolution in CF care through therapies that target the underlying CFTR defect for many patients, improving mucus properties and lung function. Better airway clearance techniques, inhaled antibiotics, and multidisciplinary care have extended life expectancy dramatically compared to earlier eras. In wound care, specialized teams and improved offloading devices have reduced amputations for many patients. These successes reinforce a central point about biofilm disease: the most effective interventions often combine microbiology with mechanics and physiology. Change the habitat, and the siege can end.

Still, it is worth lingering on the unsettling elegance of the biofilm strategy. A biofilm does not need to overpower the immune system in a straight fight. It needs to force the immune system into wasteful modes of attack, where collateral damage is high and microbial clearance is low. It needs to turn the host's urgency into the biofilm's advantage, drawing a flood of neutrophils whose enzymes weaken the very tissue that might otherwise heal over and close the niche. In that sense, chronic biofilm disease is not just infection; it is a misdirection of the body's own strength.

When a person with CF coughs until their ribs ache, or a person with diabetes watches an ulcer refuse to close week after week, the visible suffering is the end of a long microscopic argument. Cells, enzymes, oxygen gradients, strands of DNA, and sticky polymers have been negotiating the terms of coexistence. A biofilm persists not because the body is indifferent, but because the body is locked in a form of combat that cannot quite reach the enemy without injuring itself, and the microbes have learned to live inside that dilemma.

## 7.3 The Oral City-State

Dental plaque is the most familiar biofilm on Earth, and that familiarity has done it no favors. It is described as "gunk," "film," "dirt," a hygiene failure made visible. The language suggests randomness: a messy coating that ought not to be there. Yet plaque is less like dust on a shelf and more like a crowded city built on a shoreline that is flooded, drained, salted, scrubbed, and repainted several times a day.

The mouth is not a gentle habitat. Saliva gushes and slows; temperatures swing with coffee and ice water; oxygen is plentiful on the tongue but scarce in the depths of the gums; pH can drop sharply after sugar; surfaces are constantly shed and renewed. If bacteria persist here, it is not because conditions are easy. It is because the inhabitants have learned to build, cooperate, and take turns.

The first surprise is that teeth are not alive in the way skin is alive. A tooth's enamel is a hard mineral surface with no blood supply and no self-repair in the usual cellular sense. In ecological terms, it is an exposed rock. Skin can shed colonized cells. The gut can flush. The tooth simply stands there. For microbes, that stability is priceless. For the host, it is a problem that must be solved by engineering rather than regrowth: saliva, immune molecules, and mechanical cleaning.

Seconds after a tooth is cleaned, it begins to acquire a coating. Salivary proteins and glycoproteins adsorb to enamel and form what is called the acquired pellicle. This is not plaque yet; it is the stage set. The pellicle acts as a selective glue, presenting binding sites that certain bacteria recognize. It also protects enamel from acid to some extent, which is the host's interest. The pellicle is therefore a compromise surface: it makes the tooth less vulnerable, and in doing so it creates a landscape that microbes can read.

Then the first colonizers arrive.

Early colonizers are often streptococci and related organisms that can attach to pellicle components using adhesins. They do not arrive as a villainous horde. Many are common residents, tolerated by the host, sometimes even contributing to a stable oral ecosystem. But their true ecological importance lies in what they make possible. They change the local chemistry by consuming oxygen, producing acids, and secreting extracellular polymers. They also provide docking sites for other microbes through a process called coaggregation: bacteria stick to each other, not merely to the tooth. The first layer becomes living infrastructure, a biological velcro.

This is succession in the ecological sense: the community assembles in an ordered sequence because early occupants modify the environment in ways that determine who can thrive next. A clean tooth is not a neutral stage; it favors certain pioneers. Once the pioneers settle, they build conditions that favor a different set of organisms, and then another. If the environment remains balanced, the succession can stabilize into a community that coexists with the host. If conditions shift—diet, saliva flow, immune tone, cleaning habits—the same process can tilt into dysbiosis, an unhealthy community structure that drives disease.

Thinking in terms of succession makes plaque feel less moralistic. It is not simply "you didn't brush." It is "a habitat assembled, step by step, under the selection pressures of your daily life."

The second surprise is that plaque is not only a heap of cells. It is a material.

The extracellular polymeric substance, the matrix, is as central in plaque as it is on a catheter or in a chronic wound. In the mouth, a major component is polysaccharide produced from dietary sugars. These sticky polymers help bacteria adhere, clump, and resist being washed away. The matrix traps nutrients and enzymes. It slows diffusion, creating microenvironments where pH can be dramatically different from the saliva outside. It is here that the everyday act of eating can become a chemical event with consequences for enamel.

After a sugary snack, bacteria that ferment carbohydrates rapidly produce acid. If the plaque is thin and loosely organized, saliva can buffer that acid and wash it away. If the plaque is thick and structured, acid can linger. The pH at the tooth surface can fall low enough to dissolve mineral from enamel, initiating caries. The process is not a single catastrophic event. It is cumulative, the result of repeated episodes of demineralization that outpace remineralization. Plaque, in other words, is a device that can concentrate and prolong a chemical attack.

This brings the cast of characters into focus. *Streptococcus mutans* has become the emblem of tooth decay, and for good reasons: it adheres well, produces acid, tolerates acidic conditions, and can build matrix. Its virulence traits are regulated, in part, by quorum sensing—chemical signaling that allows bacteria to alter gene expression depending on population density. In *S. mutans*, a peptide-based signaling system (often summarized as the CSP pathway) influences competence, stress responses, and biofilm behavior. The important point is not the names of the molecules, but the logic: when enough neighbors are present, bacteria behave differently. They invest in communal traits that make sense only in a crowd.

Quorum sensing is sometimes described as bacteria "talking." The metaphor is imperfect, but it captures something real. Bacteria release small molecules into their surroundings; other bacteria detect these molecules and adjust behavior. Some signals are species-specific; others, such as AI-2, can be used by multiple species and act as a kind of shared chemical currency. In plaque, where many species live cheek by jowl, these signals help coordinate life in a crowded environment: when to produce matrix, when to shift metabolism, when to exchange genes, when to hunker down under stress.

Gene exchange matters because plaque is not a monoculture. Hundreds of microbial species can be present in the mouth, and they do not merely coexist; they trade. Horizontal gene transfer—the movement of genes between organisms that are not parent and offspring—can occur in biofilms where cells are in close contact and extracellular DNA is abundant. This does not mean that every dental biofilm is a laboratory of antibiotic resistance, but it does mean that the biofilm lifestyle is, in principle, a setting where genetic innovation can spread efficiently. The city has markets.

The third surprise is that periodontal disease, the chronic inflammation and tissue loss around teeth, is not simply "too many bacteria." It is a shift in community structure and in the host's response to that community.

Subgingival plaque—plaque below the gumline—lives in a special environment. It is less exposed to oxygen and more exposed to inflammatory fluid from the host. It can become a dense biofilm in a narrow crevice, where mechanical removal is difficult and the immune system's access is both close and ineffective. Here the old logic of frustrated phagocytosis returns, though in a different texture. Neutrophils stream into the gum crevice in large numbers. They release enzymes and reactive molecules. The tissue becomes inflamed.

Yet the biofilm persists.

Microbiologists have long noted that certain sets of species tend to co-occur in disease-associated plaque. A famous grouping, often called the "red complex," includes *Porphyromonas gingivalis*, *Tannerella forsythia*, and *Treponema denticola*. The names sound like a roll call from an ancient tribunal. Their importance is not that they alone cause disease, but that their presence often marks a community state associated with periodontitis. Plaque here looks less like dirt and more like political coalition: certain members appear together, and their combined activity shapes the environment.

*P. gingivalis* has been proposed as a "keystone pathogen," an organism that can have an outsized effect on the community and on host responses even if it is not numerically dominant. The keystone idea comes from ecology: remove a keystone species and an ecosystem can collapse or shift dramatically. In the mouth, the suggestion is that *P. gingivalis* can manipulate host immune signaling—particularly pathways involving complement and Toll-like receptors—in ways that reduce effective bacterial clearance while sustaining inflammation. That is an almost diabolical trick. Inflammation supplies nutrients in the form of tissue breakdown products and gingival crevicular fluid; reduced clearance allows the whole community to persist. The benefits spill over to other organisms. The host experiences chronic inflammation; the microbial community experiences a steady food supply and a guarded niche.

This is not a simple "invader" story. Periodontitis is often described as the result of a dysbiotic biofilm: a community that has shifted to a state that provokes and benefits from inflammation. The host response is part of the disease mechanism, not merely a reaction. In that sense, gum disease resembles other chronic biofilm conditions: the immune system is engaged, but the engagement is unproductive and damaging.

A concrete way to feel this is to picture the gumline not as a clean edge but as a border checkpoint. Above the gumline is an exposed surface that can be brushed and bathed in saliva. Below is a narrow, sheltered space that the host patrols with immune cells. When plaque thickens in that space, the border becomes congested. Neutrophils accumulate. Enzymes spill. The gum tissue swells and bleeds more easily. The pocket deepens, creating a larger sheltered zone for anaerobic bacteria. The city expands by changing the geography of its own border.

The mouth also offers a rare opportunity in human biology: biofilm formation can be watched, manipulated, and interrupted on a human timescale. Brush your teeth thoroughly before bed, and your tongue can feel the difference by morning. Skip brushing for a day, and the film thickens. Skip for longer, and plaque begins to mineralize into calculus (tartar), a hardened scaffold that is far harder to remove. Microbes exploit both softness and hardness: soft matrix for flexibility, hard mineral for permanence. Calculus is essentially plaque turned into stone, a public monument to microbial persistence.

There is a small historical anecdote hiding in plain sight every time a dentist discloses plaque with dye. The dyes are a reminder that for centuries, the mouth's microbial world was invisible, and therefore easy to misunderstand. Before microscopy became common, "tooth worms" were blamed for decay in many cultures. The idea sounds quaint now, but it captured a real intuition: something living seemed to be at work. When Antonie van Leeuwenhoek peered at material scraped from his own teeth in the 1670s, he described "little animals" moving vigorously. His astonishment was not only at their existence, but at their number. Plaque was no longer a vague impurity; it was a teeming society.

Modern oral microbiology has refined that shock into a map. Researchers now build plaque-like communities in saliva-conditioned flow cells, letting saliva serve as the nutrient source while bacteria coaggregate into structured, layered biofilms. Under controlled flow, the community can form recognizable architecture: towers, channels, stratified zones where oxygen and nutrients differ. The point of these models is not to mimic a mouth perfectly, but to show that plaque structure is not incidental. It emerges because the biofilm lifestyle naturally produces organization when cells attach, divide, and secrete matrix under flow.

Once plaque is seen as a city-state, familiar clinical advice becomes more interesting. Flossing is not merely about removing "stuff"; it is about disrupting a protected niche between teeth where brushing bristles rarely reach. Fluoride is not merely a chemical patch; it changes the enamel's resistance to acid and can inhibit bacterial metabolism at high local concentrations. Saliva is not merely spit; it is a buffering system, a lubricant, a transporter of antimicrobial molecules, and a habitat shaper. Dry mouth, whether from medications, radiation therapy, or autoimmune disease, is therefore not a minor inconvenience but an ecological upheaval that can accelerate caries and alter plaque composition.

Even diet can be reframed. Sugar is not simply "food for bacteria." It is a substrate that drives matrix production and acid generation, and it can shift selection toward organisms that thrive in acidic conditions. Frequent snacking keeps the plaque environment repeatedly acidified, favoring aciduric species and pushing the community along a trajectory that makes future acidification easier. The city learns to live in the fumes it creates.

There is also an emotional relief in this framing. Many people experience dental disease as personal failure, even shame. The biofilm perspective does not remove responsibility, but it replaces moral judgment with mechanics and ecology. A person with limited access to dental care, a person with reduced saliva from medication, a person with braces that create new niches, a person with depression who struggles with daily routines—all face altered selection pressures. Plaque responds to those pressures predictably, because it is an ecosystem, not a stain.

The mouth, finally, is a reminder that biofilms are not rare accidents. They are the default mode of microbial life on surfaces, including surfaces that belong to us. In a healthy mouth, plaque is not absent; it is managed. It is kept thin, disrupted, and prevented from hardening into protected structure. The difference between health and disease is often not whether microbes are present, but whether the city's growth is contained by the daily, mundane acts that keep its walls from becoming permanent. A toothbrush is, in its own quiet way, an instrument of urban planning.

Chapter 8

# Rust, Rigs, and Refineries

*Modern industry wages a relentless, multi-billion dollar chemical war against microscopic squatters that dissolve steel pipelines and drag down supertankers. Yet, in a twist of biological irony, these same sticky communities serve as the indispensable engines of our sewage treatment plants and environmental cleanup crews.*

## 8.1 The Friction Tax

A modern cargo ship looks like a clean geometric object from the quay: crisp lines, fresh paint, a name in tall letters. Under the waterline it is something else entirely. The ocean is not empty space; it is a crowded city of drifting larvae, algal spores, bacteria, and the fine dust of organic matter. Every hard surface is prime real estate. Give seawater a week and it will leave a calling card. Give it a month and it will begin to build.

The first tenants are usually microbes. They arrive invisibly, stick to the hull's paint, and begin secreting a glue-like matrix of sugars, proteins, and DNA. That matrix is often called EPS, short for extracellular polymeric substances. Think of EPS as a combined scaffold and weatherproofing: it holds cells in place, traps nutrients, and makes the surface feel less like inert metal and more like living habitat. The result may be nothing more than a slick sheen, the kind sailors call *slime*. It can be so thin that a diver has to run a gloved hand across the hull to be sure it is there.

And yet slime has weight in a ledger.

A ship moving through water pays an unavoidable cost in friction. At the scale of a tanker, most of the engine's effort is spent not in pushing water aside like a plow, but in dragging a layer of water along the hull. Water molecules "stick" to the hull at the very surface; the next layer of water sticks to the first; and a gradient of velocity forms, from nearly zero at the paint to full flow a short distance away. That near-hull region is called the boundary layer. If the hull were perfectly smooth, the boundary layer would be thinner and more orderly, and the ship would waste less energy stirring water into turbulence. Real hulls are never perfectly smooth, but engineers work hard to keep them close: coatings are formulated, surface roughness is measured, and new paints are marketed with the promise of reducing drag by a percent here, two percent there.

Biofilms, even before they become forests of barnacles and seaweed, interfere with that careful accounting. Their EPS swells with water, forming a soft, uneven micro-landscape. It is not like sandpaper, not exactly; it is more like a plush carpet laid onto the hull, flexible under flow, always rearranging itself. Flow passing over such a surface is prone to become turbulent sooner and to remain turbulent longer. In practical terms, that means the ship must burn more fuel to maintain the same speed.

This is the friction tax: a cost paid not because anything is "broken," but because something has quietly changed the physics at the interface.

One way to appreciate the stakes is to remember that the shipping industry's margins often hinge on small percentages. A large container ship can consume on the order of tens to hundreds of tonnes of fuel per day, depending on speed and conditions. Fuel is one of its dominant operating costs. If a thin biological film increases resistance by only a few percent, that becomes thousands of dollars a day for a single vessel. Multiply by weeks at sea and by the global fleet and the number becomes the kind that makes executives attentive and engineers creative.

Long before anyone knew what a biofilm was, navies understood the pain of fouling. Wooden ships sat in harbor and emerged slower, less maneuverable, and more expensive to sail. The famous 18th-century solution was copper sheathing: ships were clad below the waterline with copper plates, a technology widely adopted by the Royal Navy. Copper ions are toxic to many marine organisms, and the sheathing reduced fouling dramatically. The benefit was strategic as well as economic. A fast ship can choose when to fight and when to flee; it can escort convoys more effectively; it can outrun trouble. Speed, in other words, was a weapon, and fouling was a kind of sabotage performed by nature.

Copper was not a perfect fix. It corroded, it was costly, and it had to be installed and maintained. Modern ships mostly use specialized paints rather than copper plates, but the logic has not changed: it is cheaper to prevent organisms from settling than to haul a vessel out of the water and scrape it clean. Even so, prevention is never complete. Coatings age. Routes change. Idle time in warm ports gives colonizers a chance to establish themselves. Fouling pressure is uneven across oceans and seasons. A ship is not an object moving through a sterile laboratory tank; it is a mobile habitat with a predictable schedule, repeatedly visiting the same ecological neighborhoods.

Biofilms matter here because they are early and because they are enabling. A clean-coated hull is chemically and physically different from a slime-coated hull. The microbial layer alters surface charge, traps iron and organic particles, and creates a textured foundation where larger organisms can gain purchase. In that sense, slime is not only drag in its own right; it is often the welcome mat for the more dramatic tenants that follow. Barnacles and tubeworms are obvious and grotesque, but by the time they arrive, the hidden work has already been done.

It is tempting to imagine drag as something linear: twice the roughness, twice the cost. The boundary layer refuses such simplicity. At ship scales the flow is highly turbulent, and small changes in near-wall conditions can tip the balance between different regimes of friction. The same is true in many industrial systems where water is forced through pipes and across heat-exchange surfaces. What begins as a microscopic film can become an engineering problem because it changes how the fluid behaves.

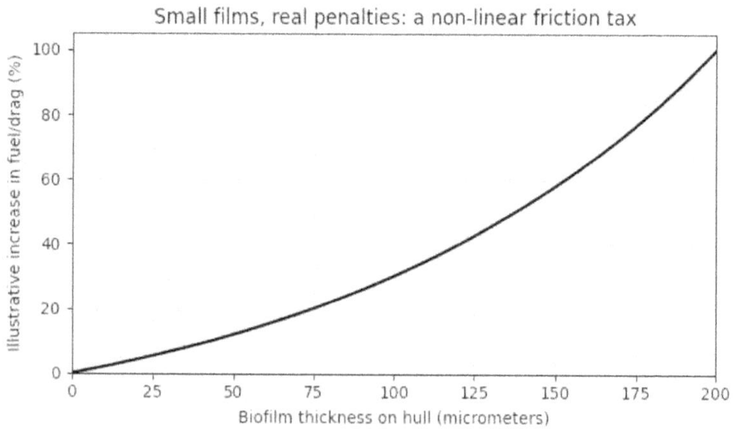


The point of a schematic curve like this is not to claim that a 100 $\mu$m film always causes a particular percentage increase. Real hulls differ, routes differ, speeds differ, and "biofilm thickness" itself is an imperfect measure because a soft gel, a patchy film, and a mineralized mat can share the same average thickness while behaving very differently in flow. The curve instead highlights a qualitative truth that frustrates planners: the early stages are not negligible. A ship can look clean to the casual eye and still be paying for the city growing on its skin.

The tax is collected in other ways too. Increased drag can force schedule changes, and schedules in shipping are not simply matters of convenience. Ports have windows. Canals have slots. Crews have contracts. A ship that arrives late can miss a berth and wait at anchor, burning fuel for generators while it waits. Shipping companies sometimes respond by increasing speed to recover lost time, and speed is expensive: the power required to push a ship through water rises steeply with velocity. A modest drag penalty can be compounded by the decision to "make up time," creating a cascade of costs that is hard to see if one only compares fuel bills from week to week.

If ships are the glamorous face of the friction tax, pipelines and industrial plumbing are its quiet bureaucracy.

A pipeline is a promise: if you put fluid in one end at a certain pressure, a predictable amount will come out the other end at the other side. That predictability makes modern industry possible, from municipal water supply to oil and gas production to food processing. Biofilms complicate the promise because they narrow the effective diameter of pipes and roughen the walls, both of which increase the pressure needed to maintain flow. This is sometimes called *biofouling* when organisms accumulate on surfaces, and *bio-clogging* when the buildup restricts flow enough to cause operational problems.

The early phase can look deceptively mild. A thin film forms on the inside of a pipe, nourished by trace organics or by the very chemicals the industry is transporting. In many systems, the flow is fast enough to discourage large organisms, and engineers might assume that "nothing can grow in there." Microbes do not need much. They can live on parts-per-million concentrations of nutrients, scavenging what passes by, and they are well adapted to life under shear stress. EPS is, among other things, a mechanical adaptation: it helps a community hold on under flowing conditions that would wash away single cells.

As the film thickens, it changes the hydraulics. The roughness promotes turbulence and increases friction losses. A pump must work harder to overcome those losses. That means more energy use and more wear. Operators may increase chemical dosing to control microbes, but dosing is a balancing act: too little does not help; too much can damage materials, create downstream problems, or violate discharge regulations. Some facilities respond with periodic pigging (sending a cleaning device down the pipe), but that interrupts operations and is only as effective as the accessibility of the system. In complex networks, there may be sections where a pig cannot pass or where flow patterns create stubborn reservoirs of biofilm.

The friction tax becomes especially vivid in heat exchangers. A heat exchanger is designed to move heat efficiently from one fluid to another. You can picture it as a radiator in reverse: instead of shedding heat to air, it may transfer heat from hot industrial fluid to cooler water, or from steam to process liquid. The performance of a heat exchanger depends on two things at once: the flow of fluids and the conductivity of the materials separating them. Biofilms attack both. They roughen surfaces and disrupt flow, but more importantly they act as insulation. EPS is mostly water, and water is not a great conductor of heat compared with metal. A biofilm layer only a fraction of a millimeter thick can form a stubborn thermal blanket. Engineers call the accumulated penalty *fouling resistance*.

The misery of fouling resistance is that it invites a familiar human response: turn the dial. Increase the temperature difference, increase flow, increase pumping power, increase steam use. Each increase costs money and often increases emissions. The plant still runs, so the problem feels less urgent than a dramatic breakdown, but the bill grows day by day. This is why fouling is often described as a hidden cost: it does not announce itself with a bang but with a slow decline in efficiency.

A concrete example comes from the world of offshore oil and gas production, where seawater is routinely used for cooling and sometimes injected into reservoirs to maintain pressure. Seawater is biologically rich and chemically complex. If it is drawn into a system and warmed, filtered, pressurized, and sent through heat exchangers and pipes, it brings microbes along for the ride. Even when biocides are used, the system can develop protected niches: dead legs with low flow, crevices near welds, places where oxygen is depleted and different microbes thrive. Operators may notice that a pump is drawing more power for the same throughput, or that a heat exchanger needs cleaning more often to maintain temperature control. Each cleaning involves labor, downtime, and safety risk. Offshore, where every intervention requires planning and often a crane, the operational costs multiply. The microbes do not need to "eat" the metal to impose this tax; they only need to build their apartments on the wrong walls.

To understand why these microbial apartments are so resilient, it helps to remember what EPS does for the community. EPS is sticky, but it is also a filter. It traps particles from the flow, including corrosion products, clays, and organic debris. In a pipe carrying water with even modest particulate load, a biofilm can become a composite material: microbial gel reinforced with trapped solids. That composite can be surprisingly tough. It can also become patchy, forming streamers and clumps that flutter in flow like aquatic weeds. Those streamers increase turbulence and can tear free, moving downstream to seed new locations, clog valves, or foul sensitive equipment.

A second reason is communication. Microbes in biofilms are not isolated individuals; they are part of a collective. Many species use chemical signaling, often called quorum sensing, to coordinate behavior based on population density. Signaling can regulate EPS production, attachment, and dispersal. In the context of industrial flow systems, this means the community can shift its physiology when it senses it has enough members to invest in a thicker matrix or to release cells to colonize downstream surfaces. The "decision" is not conscious, but it is strategic in the evolutionary sense: it helps the group persist under changing conditions.

There is also an awkward truth about shear. High flow can remove loosely attached cells and scrape away weak films. But shear can also strengthen what remains, selecting for lineages and phenotypes that produce more robust EPS or adhere more tenaciously. In other words, a system that is intermittently scoured may end up dominated by the microbial equivalent of tenacious urban planners. The same logic applies to cleaning cycles. If cleaning does not remove the entire biofilm, the survivors may be those best equipped to rebuild quickly, sometimes with a different composition than before.

Ship operators and plant engineers therefore face a common problem: they are negotiating with organisms that respond to pressure, literally and figuratively. The friction tax is not merely a mechanical inconvenience; it is a biological adaptation playing out in expensive places.

What makes the tax especially galling is how it scales. A single barnacle is not a disaster. A few micrometers of slime is not a crisis. But on a hull the area is vast, and in a refinery or municipal water system the total length of piping is enormous. The costs do not come from any one dramatic event; they come from the arithmetic of surfaces.

There is an irony here that anyone who has tried to keep a kitchen sink drain clean will recognize. The places that matter most are the places you cannot easily see. A ship's hull is underwater. The inside of a pipe is hidden. The inside of a heat exchanger is sealed. These are environments where feedback arrives late and in indirect form: an energy bill, a temperature drift, a schedule slip, a pump that seems to age too quickly. Biofilms exploit the delay. They thrive in the gap between cause and consequence.

In response, industries have developed their own forms of microbial literacy. Hull coatings are tested not only for their ability to repel larger organisms but also for how they influence the early microbial film. Some coatings aim for toxicity, discouraging settlement chemically; others aim for *foul release*, creating surfaces so slick that organisms cannot get a stable grip and are shed by motion through water. In pipes, operators monitor pressure drops, flow rates, and heat-transfer coefficients as proxies for biological buildup. They deploy biocides, adjust nutrient levels where possible, and design systems to reduce dead zones. Yet even the best strategies acknowledge a humbling fact: in wet, nutrient-bearing environments, a clean surface is an actively maintained achievement, not a default state.

The friction tax is therefore less like a fine and more like rent. If you operate in the microbial world, you pay for the privilege, either in prevention or in fuel, either in cleaning schedules or in wasted heat. The bill arrives as physics, but its origin is ecology: communities building on surfaces, turning smooth designs into rough realities, one sticky layer at a time.

## 8.2 Eating Iron

Rust is usually taught as a simple romance between iron and oxygen. Leave a nail in the rain and it turns orange; the chemistry feels almost polite, like a slow aging. Industrial failures are rarely so genteel.

When a steel pipeline develops a pinhole that grows into a leak, or when a platform's support strut shows deep pits under an apparently intact coating, the damage often looks less like weathering and more like predation: localized, opportunistic, and strangely fast.

Microbially influenced corrosion (MIC) is the name engineers give to that unsettling partnership between biology and electrochemistry. The phrase can sound modest, as if microbes merely *influence* a process that would happen anyway. In practice, they can reshape the rules. A biofilm on metal does not just sit there; it engineers a micro-environment. It changes who gets oxygen, who goes without, where acids accumulate, where protective films dissolve, where electrical currents prefer to flow. The metal becomes the ground on which a microbial city is built, and in some neighborhoods the city pays its rent in electrons stolen from the steel itself.

The story begins with something that looks harmless: a wet surface.

Steel in seawater, soil, or a pipeline never meets a uniform environment. Even without microbes, corrosion is an electrochemical process that depends on local differences. Some areas become anodic sites, where iron atoms leave the metal as ions, releasing electrons:

$$\mathrm{Fe} \rightarrow \mathrm{Fe}^{2+} + 2\mathrm{e}^{-}.$$

Elsewhere, those electrons are consumed in a cathodic reaction. In oxygenated water the cathode often reduces oxygen:

$$\mathrm{O_2} + 2\mathrm{H_2O} + 4\mathrm{e}^{-} \rightarrow 4\mathrm{OH}^{-}.$$

The anode and cathode can be millimeters apart on the same piece of steel. A pipeline is, in effect, a battery stretched for miles.

Biofilms intensify that patchiness. Their sticky matrix (EPS) is a diffusion barrier: it slows the movement of oxygen and ions. That is not merely an inconvenience. It means that two points a hair's breadth apart can inhabit different worlds. The outer face of a biofilm, washed by flowing water, may be oxygen-rich; the inner face, pressed against the metal, may be oxygen-poor or fully anaerobic. If the biofilm is uneven—thicker in one patch, thinner in another—it can create sharp oxygen gradients across the surface. Engineers call the resulting pattern a differential aeration cell. In plain terms, the better-ventilated areas tend to become cathodic, and the suffocated areas tend to become anodic. The oxygenated patch effectively encourages electron consumption; the oxygen-starved patch is left to supply electrons by dissolving iron.

This alone can drive pitting, the most treacherous kind of corrosion. Uniform corrosion is like a slow sanding down of a surface; it can be measured and anticipated. Pitting is a betrayal. A metal surface can look intact while deep cavities form beneath deposits, tubercles, or coatings. A pit concentrates stress, and a pit can punch through a wall long before the average thickness has noticeably decreased.

Microbes make pits more likely because they specialize in making micro-environments stable. The EPS matrix holds water in place, traps corrosion products, and concentrates metabolites. If microbial metabolism produces acids, those acids may linger against the metal rather than being washed away. If metabolism produces sulfide, sulfide can react with iron to form black iron sulfide deposits, altering the chemistry and electrical properties at the surface. The result is not a general acceleration of "rusting" everywhere, but an increase in the odds that a particular spot becomes a perfect anode—and stays that way long enough to carve a pit.

Among the cast of organisms associated with MIC, sulfate-reducing bacteria (SRB) have an almost legendary reputation. They are not a single species but a functional guild: microbes that can live without oxygen by using sulfate ($SO_4^{2-}$) as a terminal electron acceptor.

Sulfate is abundant in seawater and present in many subsurface and industrial waters. When SRB reduce sulfate, they produce sulfide ($H_2S$ in its familiar, rotten-egg form, or $HS^-$ depending on pH). Sulfide is chemically reactive, corrosive in its own right under many conditions, and socially useful to a biofilm because it can exclude competitors and reshape mineral deposits.

The classic explanation for why SRB accelerate corrosion has a memorable name: cathodic depolarization. It hinges on hydrogen. Even in the absence of oxygen, steel can corrode if there is a cathodic reaction available to consume electrons. One such reaction is the reduction of protons to hydrogen gas:

$$2H^+ + 2e^- \rightarrow H_2.$$

In many anaerobic corrosion scenarios, the cathodic surface becomes coated with adsorbed hydrogen atoms, which can slow further reaction. The old idea was that SRB remove this "hydrogen film" by consuming hydrogen as an energy source, thereby keeping the cathode active and allowing more iron dissolution at the anode. The metaphor is of a factory line kept moving by someone constantly clearing away the products.

That story is not entirely wrong, but it is no longer the whole story. Real biofilms are not simple, and steel surfaces are not ideal electrodes. Over the last couple of decades, corrosion scientists and microbiologists have accumulated evidence that some microbes can take electrons more directly from metals, using forms of extracellular electron transfer (EET). The phrase sounds arcane, but the concept is easy to picture: instead of waiting for hydrogen to form as an intermediate, a microbe may tap into the metal's electrons through conductive pathways. Those pathways can be specialized proteins on the cell surface, conductive appendages, or even networks of corrosion products that behave like semiconductors. The metal becomes part of the microbe's electrical landscape.

If that is true for a given community in a given environment, the steel is not merely providing a surface; it is providing fuel in a very literal sense. The microbe's metabolism is coupled to the oxidation of iron. The iron dissolves because the electrons have somewhere biologically useful to go.

It also helps explain a practical observation that has long frustrated simplistic models: MIC is often community-driven. A single SRB strain grown in a laboratory bottle does not always reproduce the severity and pattern of corrosion found in the field. Pipelines, ballast tanks, and buried structures host consortia: aerobes and anaerobes layered within the same biofilm, iron oxidizers and iron reducers cycling metals back and forth, organisms that produce EPS, organisms that degrade it, organisms that ferment organics into acids, organisms that consume the acids. The electrochemistry of steel is braided into a living food web.

A vivid way to see that braiding is to watch what happens when steel first enters a natural environment. The earliest colonizers are often not SRB, because SRB prefer low-oxygen conditions and newly exposed steel in moving water may be fairly oxygenated. Iron-oxidizing bacteria (FeOB) are plausible pioneers. They gain energy by oxidizing ferrous iron ($Fe^{2+}$) to ferric iron ($Fe^{3+}$) under microaerobic conditions, where there is a little oxygen but not too much. Their metabolic niche is narrow, but steel surfaces and their near-boundary layers can provide it. Many FeOB build delicate iron-oxide structures—stalks and sheaths—that are as much architecture as waste product. Those structures create a three-dimensional scaffold that traps other cells and particles, turning a bare surface into a textured habitat. Once that habitat exists, anaerobic pockets can form more readily, and SRB can move in.

The result is ecological succession with an electrochemical punchline. Early colonizers change the surface and the local chemistry; later colonizers exploit the new conditions; the biofilm thickens; gradients sharpen; corrosion localizes.

Engineers first confronted the industrial consequences of SRB long before they could describe the organisms in modern terms. In the early twentieth century, as steel pipelines and storage tanks proliferated, the pattern of failures in certain soils and waters seemed too aggressive to be explained by simple chemistry. There is a famous historical thread linking "anaerobic bacteria" to corrosion in oil fields, where black deposits and the smell of sulfide were repeatedly associated with damaged steel. The organisms involved were eventually identified and cultivated, and the label SRB entered industrial vocabulary. Even today, field technicians will talk about "souring" and "sulfides" with a kind of wary familiarity, because those smells can signal both operational problems and corrosion risk.

A concrete modern example comes from water distribution systems. Drinking water utilities fight corrosion for public health reasons as well as for asset longevity: corroding pipes can release metals, create discolored water, and provide niches for microbial regrowth. Biofilms in distribution pipes are normal; the question is what kind of biofilm, and on what material. Unlined iron surfaces tend to develop tubercles: rough mounds of corrosion products and biofilm that can shelter anoxic zones inside. Within and beneath these deposits, oxygen is scarce and sulfate reducers, iron reducers, and other anaerobes may find refuge even though the bulk water remains oxygenated. A resident might notice intermittent "brown water," a musty smell, or changes after a main break flushes deposits downstream. From the utility's perspective, the problem is not merely aesthetic. The deposits change hydraulics, increase energy costs for pumping, and complicate disinfection. In extreme cases they contribute to leaks and breaks that damage streets and homes. The boundary between microbiology and civil infrastructure becomes very thin.

The distinctive cruelty of MIC is that it can create damage out of proportion to the amount of material involved. A pinhole leak in a pipeline can release thousands of gallons of product before it is detected. In an offshore setting, a small failure can become an environmental disaster because of the difficulty of access and repair. In a refinery, a localized failure in a heat exchanger tube can force an unscheduled shutdown, with costs that dwarf the price of the metal itself. Corrosion is not merely the loss of steel; it is the loss of trust in systems designed to be dependable.

The physics and chemistry of MIC can be intimidating, but the essential mechanisms can be held in mind without equations. A biofilm acts like a blanket and a baffle. It blankets the surface by slowing diffusion and trapping chemicals. It baffles the system by creating patchiness: different micro-environments side by side. In those micro-environments, microbes pursue their own survival strategies, and some of those strategies involve shuttling electrons in ways that keep corrosion reactions running. The metal's fate depends on where electrons are produced, where they are consumed, and how steadily the conditions are maintained that favor one spot becoming an anode.

The "heist of electrons" metaphor is more than rhetorical flourish. Microbial metabolism is, at its core, a set of redox reactions: taking electrons from one place and delivering them to another. Humans do the same thing with power stations and batteries. Microbes simply do it at microscopic scales and with astonishing chemical flexibility. When a biofilm sits on steel in a sulfate-rich environment, the steel can become the electron donor of convenience.

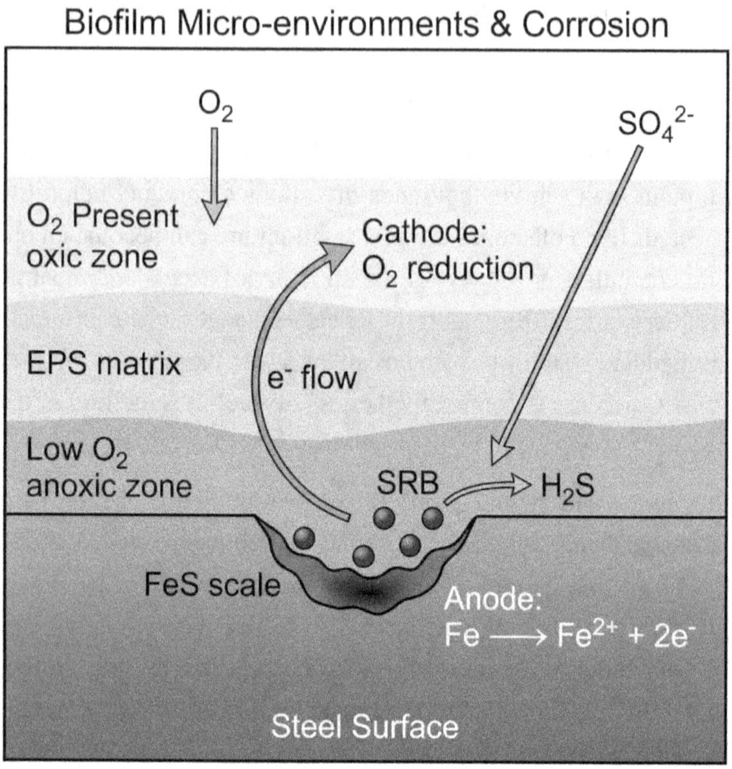


A diagram like this is necessarily a simplification, but it captures the logic of how a pit can become a persistent neighborhood. The thick biofilm layer slows oxygen diffusion to the metal. Near the top, oxygen is available; near the bottom, it is scarce. That separation allows anodic and cathodic regions to coexist on the same surface. If SRB establish themselves in the anaerobic zone, sulfate can serve as an electron sink for their metabolism, and sulfide can accumulate. Where sulfide meets iron, black crusts of iron sulfide can form, which can further change local electrochemistry and help maintain the very gradients that made the niche attractive.

The most uncomfortable feature of MIC is that it can coexist with the very systems meant to prevent it. Protective coatings can delaminate locally, creating a crevice where a biofilm becomes sheltered from flow and biocide. Cathodic protection systems, designed to push the steel's potential into a range where corrosion should be suppressed, can be complicated by biofilms that create local chemistry different from bulk assumptions. Even the act of cleaning can create fresh surfaces and fresh opportunities for colonization if it is not paired with a strategy that changes the underlying ecology.

Because MIC is so often localized, its early detection is hard. A pipeline may lose a tiny fraction of wall thickness in a small area while the rest remains healthy. A routine inspection that samples only certain points can miss the worst site. Operators therefore rely on a mix of approaches: corrosion coupons (small test pieces inserted into the system), electrical resistance probes, fluid sampling for microbial markers, and intelligent pigging that maps wall thickness. Yet the biofilm is always one step ahead in its ability to hide: it is literally a covering.

When the damage finally becomes visible, it often looks like a secret excavation. Peel back a deposit and the steel beneath may be cratered, with a blackened smell. The pit is not just a hole; it is a record of micro-environments held in place for months or years. A human engineer might see a defect; a microbe sees a stable niche.

One of the more sobering lessons from MIC research is that "microbe-free" is rarely a realistic goal in large wet systems. The better aim is control: shaping conditions so that the microbial community that inevitably forms is less likely to create aggressive electrochemical neighborhoods. That might mean limiting sulfate where feasible, reducing stagnation, controlling nutrients, choosing materials and linings that discourage thick deposits, and designing for cleanability. It may also mean respecting that biofilms are not passive dirt. They are organized communities that respond to stress, share resources, and sometimes exchange electrons in ways that make steel an attractive target.

In that sense, the phrase *eating iron* is not mere sensationalism. For some microbes in some settings, metallic iron can function as an electron donor, part of the energy economy of a biofilm. The steel does not simply succumb to its environment; it becomes entangled in a metabolism. The orange rust we recognize is only one possible outcome. Sometimes the more telling color is black, and the more telling sign is not a uniform stain but a single pit that deepens quietly until the metal's promise of solidity gives way.

## 8.3 The Billion-Dollar Biofilter

A city's most intimate infrastructure is also its least discussed. Drinking water is ceremonial: it has fountains, branded bottles, arguments over taste. Wastewater is the opposite. It is a daily, collective act of disappearance. Pull a lever, press a button, and what your body discards slips into an underground network with the expectation that it will be dealt with, quietly and safely, far from kitchens and parks.

That expectation is not a law of nature. It is an engineering achievement built on biology.

In the industrial settings where biofilms become villains—a ship slowed by slime, a pipeline perforated by pits—the moral feels straightforward: keep microbes off valuable surfaces. Yet the same microbial talent for sticking, coordinating, and thriving in gradients becomes indispensable when the problem is not protecting metal but protecting a river. Wastewater treatment plants are, in a deep sense, cities built for microbial cities. They take a chaotic flow of organic matter, nutrients, and pathogens and turn it into water that can be returned to the environment without poisoning it. They do it at scales that would seem absurd if you had to do it with chemistry alone.

The central trick is simple to state and hard to perfect: keep a lot of helpful microbes in close contact with a lot of dirty water, and give them enough oxygen (or other electron acceptors) to finish the job. Biofilms are the solution because they are a way of packing microbial metabolism into a stable, high-density form. A free-floating bacterium in a rushing stream has little leverage. A bacterium embedded in a matrix, attached to a surface, living in a structured community, can persist through changing flow, feast when food arrives, and endure when it does not.

The earliest large-scale wastewater systems were not built with modern microbiology in mind. The "sewage farms" and early filtration beds of the nineteenth century were as much agricultural intuition as science: soil and gravel can cleanse water. Where they worked, they did so because they cultivated microbial life on surfaces. Water percolating through sand is not purified by the sand's moral virtue; it is transformed by the thin living films that coat each grain. Those films are biofilms, whether the designers knew the word or not.

One of the oldest engineered biofilm systems still recognizable today is the trickling filter. The name sounds quaint, but the concept is muscular. Wastewater is distributed over a bed of rocks, slag, or specially designed plastic media. It trickles down in thin sheets, bathing surfaces that are coated in microbial growth. Air circulates through the void spaces, supplying oxygen. Organic matter in the wastewater is absorbed and metabolized by the biofilm. Pieces of biofilm slough off as they thicken and are carried away to be settled out downstream.

The trickling filter does not look like a high-tech device. It looks like a landscaped ruin: a circular bed, a rotating distributor arm, water sprayed like a lazy fountain. Yet it embodies a profound engineering insight: you can turn the surfaces of inert media into a biochemical factory without having to pump enormous amounts of air into a mixed tank. The oxygen arrives through the structure itself.

This kind of biofilm engineering is, at heart, an argument about control. In the unwanted biofilms of ships and pipelines, microbes pick the surface and set the terms. In treatment plants, humans pick the surface and set the terms, then invite the microbes to move in.

Activated sludge, the workhorse of many modern wastewater plants, makes the invitation in a different way. Instead of asking microbes to live on fixed surfaces, it encourages them to form flocs: suspended clumps of bacteria, protozoa, and extracellular matrix that behave like floating biofilms. Wastewater is mixed with this living suspension in aeration tanks; oxygen is supplied, and microbes consume organic matter and nutrients. The flocs are then allowed to settle in clarifiers, separating clean water from the biomass, which is partly recycled back to the aeration tank.

To the naked eye, activated sludge looks like brown water that settles into a soft blanket at the bottom of a settling basin. Under a microscope, the flocs are bustling metropolises: bacteria embedded in EPS, filaments acting as scaffolding, grazers feeding at the edges, channels allowing water to flow through. A floc is not just a random clump. It has structure and function. It has neighborhoods. It has, in its own way, hygiene: predators and competition help suppress some undesirable organisms, while stable conditions allow beneficial communities to persist.

The very features that make biofilms hard to kill in an industrial pipe—their matrix, their density, their ability to shelter slow-growing cells—become advantages when the goal is steady treatment. A biofilm retains biomass. It does not wash out easily, so the system can handle fluctuations: a morning surge, an industrial discharge, a heavy rain that dilutes sewage but increases flow. Biofilms also create gradients, and gradients are not merely tolerated; they are exploited.

Consider nitrogen, a nutrient essential for life and disastrous in excess. Too much nitrogen in rivers and coastal waters can drive algal blooms and oxygen depletion, harming fisheries and ecosystems. Wastewater contains nitrogen in forms like ammonia ($NH_3$ or $NH_4^+$), and removing it requires a sequence of microbial transformations.

One key step is nitrification: the oxidation of ammonia to nitrite ($NO_2^-$) and then to nitrate ($NO_3^-$). This process is carried out by specialized microbes that grow slowly and demand oxygen. Another key step is denitrification: the reduction of nitrate to nitrogen gas ($N_2$), which returns harmlessly to the atmosphere. Denitrification occurs under low-oxygen or anoxic conditions and requires an electron donor, often organic carbon.

If that sounds like conflicting requirements, it is. Engineers once had to build separate environments: an aerated zone for nitrification, an anoxic zone for denitrification, with careful recycling flows between them. Biofilms offer a more compact solution because they can contain both environments at once. Oxygen can penetrate only so far into the biofilm. The outer layer may be oxygen-rich and ideal for nitrifiers; deeper layers may be oxygen-poor and ideal for denitrifiers. The same piece of plastic media can thus host a vertical stack of metabolisms, each living off the gradients created by diffusion and consumption.

This is one reason moving bed biofilm reactors (MBBRs) have become prominent in modern wastewater treatment. An MBBR uses small plastic carriers, often shaped like spoked wheels or honeycombs, that tumble freely in an aerated tank. Biofilm grows on the carriers. Because the carriers are retained in the tank by screens, the biomass is retained without relying on settling properties in the same way activated sludge does. In practice, this can make the system more robust and allow treatment in a smaller footprint. The carriers provide an enormous surface area, meaning that a relatively compact tank can host a great deal of microbial activity.

The MBBR also makes visible a subtle fact about biofilms: they are not necessarily slimy carpets. In a well-run reactor, biofilm can be a thin, controlled layer, repeatedly sheared by collisions between carriers and the turbulence of aeration. This controlled abrasion keeps the biofilm from becoming too thick and anaerobic everywhere. It prevents excessive sloughing events. It is, in effect, city zoning enforced by fluid dynamics.

The underlying economic logic is easy to miss because it is expressed as negative space: costs avoided. Wastewater treatment prevents disease outbreaks by breaking fecal-oral transmission routes. It reduces eutrophication, protecting lakes and estuaries. It allows cities to concentrate populations without turning rivers into open sewers. These benefits are so fundamental that they vanish into normalcy, which is why treatment plants are often built on the outskirts, out of sight and out of mind.

A historical reminder of what is at stake comes from the era before effective sewage treatment and safe water supplies, when cities repeatedly suffered cholera and typhoid outbreaks. John Snow's famous 1854 cholera investigation in London helped reveal that contaminated water could spread disease, but identifying a cause did not automatically create the infrastructure to fix it. The transformation required sewers, treatment, and the social willingness to invest in systems that deliver benefits mostly by preventing catastrophes. Wastewater treatment is part of that transformation. It is a public health technology as much as an environmental one, and biofilms are among its key working parts.

Biofilms also help with a problem that is both older and more modern than sewage: the management of industrial gases and odors. Hydrogen sulfide ($H_2S$) is a classic example. It is produced by sulfate-reducing bacteria in sewers and sludge-handling systems, especially where water becomes stagnant or oxygen is depleted. $H_2S$ is toxic at high concentrations, corrosive to concrete and metals, and notorious for its smell. Anyone who has walked near an overloaded sewer line on a hot day has encountered the warning.

Chemical scrubbing can remove $H_2S$, but it consumes reagents and generates waste streams. Biofilters offer a biological alternative. In a biofilter, contaminated air is passed through a packed bed—often compost, wood chips, or engineered media—coated with a biofilm of sulfur-oxidizing microbes. These organisms use $H_2S$ as an energy source, converting it to elemental sulfur or sulfate. The gas phase becomes cleaner because the pollutant becomes microbial food.

What makes this work is not just the existence of sulfur-oxidizing bacteria, but the spatial arrangement. The biofilm sits at the interface where gas meets moisture. $H_2S$ dissolves into the thin water layer around the biofilm, then diffuses into the cells. Oxygen diffuses in as well. The community is positioned exactly where gradients are steepest and where the chemistry is most favorable. Again, biofilms thrive on boundaries, and engineered systems put them on the boundaries that matter.

In some industrial designs, the biology can even be pushed toward a kind of recovery rather than mere removal. Sulfur compounds are not only pollutants; they are potential products. Under carefully managed conditions, sulfur-oxidizing biofilms can generate concentrated sulfuric acid, turning a smelly hazard into a commodity chemical. That is not a universal or trivial outcome; it requires careful control of pH, oxygen supply, and mass transfer. But it illustrates a deeper point: a biofilm can be treated as a catalyst bed, not just as a nuisance layer. Its "waste" can be a feedstock, and its metabolism can be harnessed with the same seriousness one would apply to a chemical reactor.

This is the core reversal: the same microbial strategies that make MIC so damaging—forming protected communities, creating steep chemical gradients, exchanging electrons—can be turned toward protection rather than destruction. In wastewater treatment, the metal is not the electron donor; the waste is. Organic carbon and reduced nitrogen and sulfur compounds are energetic molecules. Biofilms harvest that energy and, in doing so, transform pollutants into less harmful forms: carbon into carbon dioxide and biomass; ammonia into nitrate and then nitrogen gas; sulfide into sulfate; countless trace organics into simpler molecules.

Even the "stickiness" that makes biofilms hard to remove becomes a design feature. A trickling filter relies on attachment. An MBBR relies on attachment. Denitrifying filters, used in some treatment trains, rely on attachment: water flows through media, biofilms remove nitrate. The attachment gives the process stability. Without it, every change in flow would carry away the workers.

There are, of course, costs and complications. Biofilm systems can clog if solids accumulate faster than they are broken down. They can develop channeling, where water takes preferential paths and bypasses the biofilm. They can slough unpredictably. They can emit nitrous oxide, a potent greenhouse gas, if nitrification and denitrification are poorly balanced. None of this is magic. Biofilm engineering is a craft of constraint and feedback: providing enough oxygen but not too much, enough carbon but not too much, enough shear to control thickness but not enough to strip the system bare.

Yet it is hard to overstate how extraordinary it is that cities can process their own waste continuously, day after day, using living communities as core machinery. A treatment plant is a kind of ecological organ, metabolizing on behalf of a population. Its biofilms are not incidental; they are the tissue.

There is also a social dimension to the "billion-dollar biofilter." Wastewater infrastructure is expensive to build and maintain, and it is often politically invisible until it fails. When it fails, the consequences can be immediate and intimate: beach closures, fish kills, basement backups, boil-water notices, a sudden return of smells and pathogens that modern cities have trained themselves to forget. Investment in treatment is therefore a kind of civic patience, paying now for a benefit that looks like nothing happening.

Biofilms make that patience pay off because they are efficient at doing what chemistry would struggle to do at scale: selectively breaking down a vast variety of organic molecules under varying conditions, self-replicating when food is abundant, and adjusting their composition to the wastes a city produces. A manufacturing plant that changes its discharge can, over time, reshape the microbial community downstream. A city that implements industrial pretreatment, reduces phosphorus inputs, or changes disinfection practices can alter the ecology of its own treatment process. The microbial city is responsive, not static.

There is a final irony worth holding onto. The sewage that arrives at a plant is a mixture of human biology and human culture: what we eat, what medicines we take, what cleaners we use, how much water we flush. The biofilm that treats it is an evolving record of those choices. A biofilter is not just a machine; it is a living archive of a city's habits, written in metabolism rather than ink, turning what we would rather not think about into water that can safely rejoin the world.

Chapter 9

# Dismantling the City

*Biofilms are not eternal fortresses; they possess innate mechanisms for self-dissolution to colonize new frontiers. By decoding the chemical signals that trigger this mass exodus or silence their coordination, scientists are forging weapons that do not merely kill bacteria but dismantle their civilizations.*

## 9.1 The Great Exodus

A thriving biofilm looks permanent. Under a microscope it can resemble a skyline: rounded towers, thick neighborhoods, and channels that feel almost like streets. Yet permanence is not the point. For a microbe, settling down is a tactic, not a destiny. Biofilms are built for a particular phase of life: holding on when conditions are good enough to justify the cost of construction, and letting go when the same architecture becomes a liability. The most dramatic moment in that life cycle is dispersal, when a fraction of the population abruptly abandons the structure and returns to the free-swimming, planktonic state.

Dispersal is often imagined as simple erosion: bits of slime torn off by a current, flakes sloughing away as the biofilm ages. That happens, but it misses what makes biofilm cities so formidable and so frustrating in medicine and industry. Many biofilms do not merely crumble; they *choose* to dismantle themselves. They trigger internal demolition, liquefy parts of their own matrix, and release a wave of cells prepared for travel. The effect can look like a seeding event: a mature tower shedding motile emigrants that can colonize downstream surfaces, spread through tissues, or repopulate a cleaned pipe.

The logic is bleakly practical. Deep in a thick biofilm, oxygen runs out. Nutrients become scarce. Waste products and acids accumulate. Even if the surface layers keep growing, the interior can become a stressed, crowded basement. Cells in those depths have a decision to make: endure, die, or leave. Dispersal is the gamble that leaving will be safer than staying.

Biofilm dispersal has another motive, less obvious but just as important: unpredictability. A biofilm is an investment in one location. If that location becomes hostile—a sudden flood of disinfectant, a pulse of antibiotics, a drying event, a hungry immune cell—a population that has already produced travelers has insurance. A mature biofilm is not only a fortress; it is also a launchpad.

To leave, though, microbes must solve two problems that are, for them, engineering problems. First, they must weaken the extracellular polymeric substances (EPS): the sticky matrix of polysaccharides, proteins, lipids, and extracellular DNA that holds the city together. Second, they must change their own bodies and behavior: a cell adapted to life in a dense community must become a cell adapted to motion, exposure, and solitary survival. Dispersal is not simply a door opening. It is a coordinated conversion.

The easiest way to picture the conversion is as a switch between two lifestyles. One lifestyle is adhesive: surface proteins and pili that grip, secretion systems that lay down glue, slow growth tuned to limited oxygen, and metabolic pathways that tolerate gradients and waste. The other lifestyle is exploratory: flagella (or other motility structures) turned back on, surface adhesins turned down, stress responses reconfigured, and often a reawakening of rapid growth. These are not moral choices, but gene-expression programs. The cell is not deciding in a conscious sense, yet it is changing in a purposeful direction.

A central character in this story is a small intracellular messenger molecule called cyclic di-GMP (often written c-di-GMP). It functions as a kind of molecular vote-counting system. When c-di-GMP levels are high, many bacteria commit to the sticky, sedentary life: they produce matrix components, reduce motility, and behave like residents. When c-di-GMP levels fall, the balance shifts: motility genes and dispersal behaviors become more likely, and the biofilm phenotype loosens. The molecule is simple; its consequences are not. Dozens of enzymes may build it or degrade it, each sensitive to different environmental cues. A biofilm is therefore not governed by a single lever but by a dashboard of sensors and feedback loops that converge on the same outcome: stay put, or get ready to move.

What cues push the dashboard toward departure? Some are mundane. Starvation is an obvious trigger: when growth slows and competition increases, dispersal can become attractive for at least a subset of the population. Shifts in oxygen are another. In thick biofilms, oxygen gradients are steep; a cell can be only micrometers away from air yet live as if it were underground. When oxygen is scarce, bacteria may use alternative electron acceptors, producing different metabolic by-products. Those by-products can become signals themselves.

One of the most striking examples is nitric oxide (NO). In human physiology, NO is famous as a signaling molecule involved in blood vessels and immune responses; it is also an industrial pollutant and a reactive gas. In microbial biofilms, low concentrations of NO can serve as a cue that encourages dispersal. The surprising part is the dose: at the right low levels, NO does not necessarily kill cells. Instead, it nudges the internal circuitry that lowers c-di-GMP and shifts the population toward motility and detachment. That discovery, first worked out in detail in *Pseudomonas aeruginosa* biofilms by Nicolas Barraud and colleagues in the mid-2000s, changed how many researchers thought about dispersal. It suggested that a biofilm could be coaxed to take itself apart, not merely blasted into rubble.

The NO story also illustrates a theme that repeats across microbiology: signals are often repurposed toxins. A molecule that is dangerous at one concentration can be informative at another. For a biofilm, NO can mean something like, *conditions are shifting; respiration is changing; it may be time to prepare to leave.* That interpretation fits with the ecology. In low-oxygen environments, bacteria may produce NO as part of denitrification pathways. If NO begins to accumulate, it can reflect the constraints of anaerobic metabolism and diffusion limits in a thick community. A biofilm that has grown too dense may be warning itself that it is approaching a tipping point.

Signals are only the first half of the exodus. The second half is demolition. The biofilm matrix is not a passive slime; it is a composite material. Extracellular DNA can behave like rebar, giving tensile strength. Polysaccharides form gels that trap water and ions. Proteins cross-link components and can act like scaffolding. To loosen the structure quickly, microbes deploy enzymes that cut these components.

DNases are among the most intuitive: enzymes that digest extracellular DNA. If eDNA is part of the structural skeleton, DNase is a pair of wire cutters. Proteases trim matrix proteins. Glycoside hydrolases clip polysaccharide chains. Some bacteria produce surfactants—soap-like molecules—that change surface tension and help detach clumps. Often these tools work together: a biofilm may soften by enzymatic clipping, then detach by fluid shear, like a building weakened by internal demolition and then toppled by a strong wind.

There is a revealing paradox here. The same species that spends energy secreting EPS also spends energy producing enzymes to degrade it. That makes sense only if the matrix is considered an instrument, not a home. A biofilm is built with an exit strategy.

A concrete example comes from chronic infections in cystic fibrosis (CF), where thick mucus and altered airway physiology create an environment in which *Pseudomonas aeruginosa* can form persistent biofilm-like aggregates. Clinicians have long recognized that these infections can behave in a punctuated way: periods of relative stability interrupted by exacerbations, sometimes associated with increases in bacterial load and inflammation. While the clinical story is more complex than any single mechanism, the biology of dispersal offers one plausible bridge between microscopic events and macroscopic consequences. A treatment that merely shakes loose biofilm bacteria without killing them risks turning a contained, tolerant population into a mobile one. In other words, dispersal is not automatically a victory for the host. It is a change in the battle's geometry. Therapies inspired by dispersal signals, such as low-dose NO approaches studied in CF contexts, have therefore been framed not as replacements for antibiotics but as adjuncts: dismantle the protected structure and then apply killing pressure while the cells are exposed.

That caution is not hypothetical; it is rooted in the natural purpose of dispersal. When bacteria leave a biofilm, they often do so as primed colonizers. Some disperse as single swimmers. Others disperse as small clumps—tiny caravans that retain some protection and can establish new biofilms rapidly. In plumbing systems, this can translate into the sudden reappearance of contamination downstream even after a local cleaning. In medical settings, it can mean the spread from one niche to another. An exodus is a survival strategy, and survival strategies rarely align neatly with human goals.

Dispersal can also be triggered by small molecules that bacteria make themselves, sometimes in ways that look almost like communication across generations. One of the most discussed examples came from work by Illana Kolodkin-Gal and colleagues around 2010, showing that certain D-amino acids (mirror-image forms of amino acids) could promote disassembly of *Bacillus subtilis* biofilms. Amino acids are usually thought of as nutrients and building blocks. Their D-forms are less common in proteins but are used in bacterial cell walls and can have signaling roles. The idea that a mixture of D-amino acids could act as a cue for biofilm breakdown suggested that dismantling might be wired into development: a mature community generates chemical conditions that favor its own dispersal, perhaps as a way to reset the habitat or spread to new ones.

There is a historical echo here, and it is instructive. Long before molecular biology could measure c-di-GMP or visualize gene expression in microcolonies, microbiologists noticed that bacterial cultures did not always grow smoothly. They changed texture. They formed films, then shed them. In the nineteenth century, when Louis Pasteur and contemporaries were wrestling with fermentation and contamination, they repeatedly encountered the practical reality that microbes were not simply suspended in liquid; they adhered, formed scums, and then broke apart. Brewers and vintners learned by experience that a vessel could be scrubbed and still become fouled again, as if the contamination had a talent for returning. Today, it is easier to name the culprit: dispersal cells escaping a protected niche and reseeding the system.

The metaphor of a city helps because it keeps attention on infrastructure. Dispersal is not only about the inhabitants leaving; it is about the roads and walls being temporarily dismantled. In many species, lowering c-di-GMP reduces production of matrix polysaccharides and adhesins while increasing motility. That internal shift is often accompanied by outward action: enzymes cut the matrix, surfactants loosen connections, and the physical architecture becomes porous. Fluids can then penetrate more easily, carrying away cells and fragments.

The physical form of dispersal depends on the biofilm's environment. In a stream, the shear forces are strong and detachment can be dramatic. In a wound, the fluid flows are different; dispersal may mean migration across tissue surfaces or into deeper pockets. In a catheter or a dental plaque layer, the mechanical forces arise from intermittent flow and from the host's attempts to remove the biofilm. A toothbrush, for example, is a dispersal tool, but an imperfect one: it can break up plaque and remove much of it, yet it can also release bacterial cells into saliva and into crevices. That is one reason dental hygiene is a repeated action rather than a one-time fix. Biofilms are built to recover, and dispersal is part of recovery as well as part of spread.

Enzymatic self-destruction also explains why biofilms can sometimes collapse in a way that feels sudden. A mature biofilm may look stable for days, then within hours become ragged, hollowed, and porous. If enough cells switch into a dispersal program at once, they can collectively change the chemistry of the matrix. Cutting the matrix is a cooperative act: if one cell secretes a hydrolase, it benefits nearby cells too. That makes dispersal vulnerable to cheaters—cells that take advantage without contributing—but it also makes it capable of rapid, community-level change once a threshold is crossed. The same principle underlies the abruptness of many group phenomena in biology: below the threshold, nothing seems to happen; above it, everything moves.

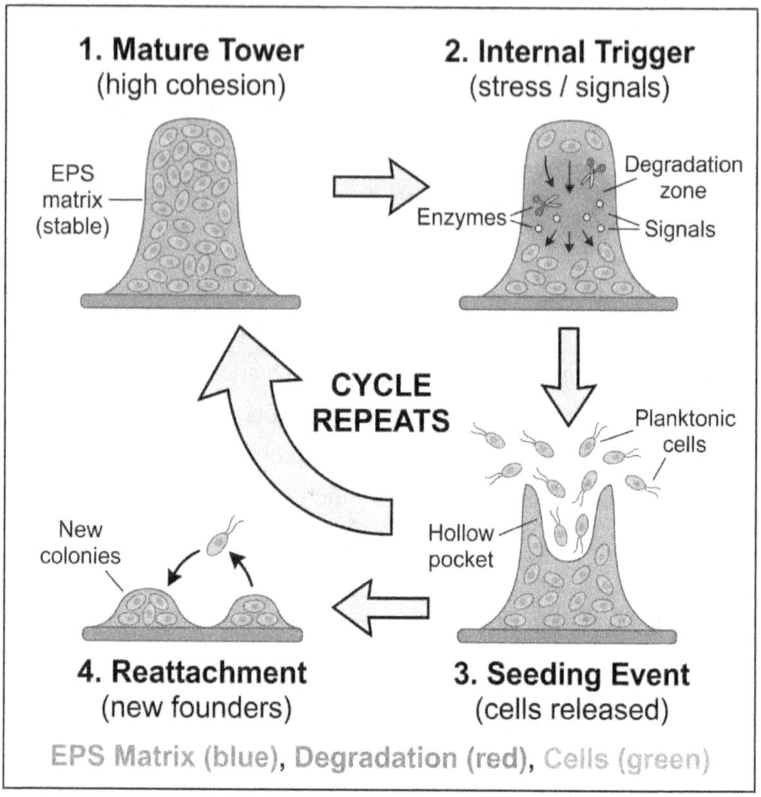


Seen this way, dispersal resembles an emergency drill that is also a reproductive strategy. It is emergency because it is often triggered by stress: starvation, oxygen limitation, toxic buildup, immune attack, disinfectants, or competitive pressure from other microbes. It is reproductive because it increases the range of the lineage. A biofilm that never disperses is a lineage that bets everything on one surface.

The subtler question is why dispersal is frequently partial. Why not dissolve the entire biofilm and flee? Part of the answer is risk management. A mature biofilm may be living in a patchwork of conditions. The top can still be well-fed and well-oxygenated. The bottom can be starving. In that situation, it makes sense for the interior to release emigrants while the surface layers continue to exploit the remaining resources. There is also safety in keeping a base camp. Even if emigrants fail to find a new home, the remaining structure can persist and perhaps rebuild. In ecology, this resembles the "seed bank" strategy of plants, where some seeds stay dormant while others germinate. In microbial terms, different phenotypes hedge against the unpredictable.

This diversity is not only behavioral; it is physiological. Dispersed cells are not necessarily identical to ordinary planktonic cells grown in a flask. They may carry a biofilm memory: altered gene expression, stress responses, and metabolic traits shaped by time spent in gradients. Some disperse as highly motile individuals. Others leave as aggregates that preserve local cooperation. Some may be primed for virulence factor expression; others may be primed for rapid growth. The exodus is therefore not a single event but a portfolio of departures.

There is a useful way to connect this to everyday experience. Consider the smell that rises when a neglected drain is disturbed. The odor is not merely a chemical nuisance; it is a reminder that the drain has become an ecosystem. When hot water surges through, when detergent is poured in, when a brush scrapes the walls, the biofilm is not only removed; it is *rearranged*. Cells and fragments can be released into the flow, traveling to new bends and joints where the surface is rougher or the water stagnates. In domestic settings this mostly means inconvenience. In hospitals it can mean something else entirely: a biofilm in plumbing can become a reservoir for opportunistic pathogens, and disturbances can release cells into aerosols or into water used for patient care. The "cleaning" moment can paradoxically be a dispersal moment.

All of this helps explain why dismantling biofilms in practice is rarely as simple as "remove the slime." If a biofilm is forced to disperse, the result can be a shower of exposed cells, and exposed cells are often easier to kill. But exposed cells are also more mobile and more capable of spreading. The safe version of dispersal, from a human point of view, is therefore constrained: induce dismantling while simultaneously applying a killing or capture step. In pipes, that might mean flushing with biocide at the moment of detachment. In wounds, it might mean pairing matrix-disrupting approaches with antimicrobials and debridement. In lungs, it might mean an adjunct that loosens aggregates paired with antibiotics and host clearance mechanisms.

The reason biofilm dispersal remains so fascinating is that it forces a redefinition of what it means to attack microbes. Killing is one strategy; changing their state is another. A biofilm is not only a collection of organisms but a temporary body plan: cells plus infrastructure plus coordinated behavior. Dispersal is the moment that body plan is intentionally abandoned. When the matrix softens, when c-di-GMP drops, when flagella reappear and cells begin to swim, the city turns itself into travelers. The walls do not fall because they are weak; they fall because, for some fraction of the inhabitants, it is time to leave.

## 9.2 The Communication Blackout

A biofilm is not only sticky; it is coordinated. That coordination is easy to underestimate because it rarely looks like anything we would call a plan. No one is issuing orders. Yet a mature biofilm behaves differently from a crowd of bacteria drifting past one another in water. It turns on certain genes at high density and turns others off. It shares costly public goods—enzymes, siderophores that scavenge iron, matrix polymers that hold water—and it times their production so that the benefits fall on neighbors likely to return the favor. It can become, in the most practical sense, a society.

One of the best-understood ways bacteria coordinate is by chemical signaling called quorum sensing. The phrase sounds like politics, but the mechanism is closer to smell in a crowded room. Each cell releases small signal molecules into its surroundings. When there are only a few cells, those molecules diffuse away and remain too dilute to matter. As the population grows, the molecules accumulate. Once the concentration crosses a threshold, receptors inside or on the surface of bacteria detect the signal and switch gene expression. The key point is not mystical: the signal concentration acts as a proxy for local cell density and confinement. High signal means, *many of us are nearby, and diffusion is limited.* That is precisely the information a biofilm cell needs before it spends energy on group behaviors.

Different bacteria speak different dialects. Many Gram-negative species use acyl-homoserine lactones (AHLs), small molecules with a lactone ring and a variable acyl tail. Many Gram-positive species use short peptides. Some signals, such as autoinducer-2 (AI-2), appear in a wide range of bacteria and are often described as a more universal tongue, though the reality is messier than the label implies. A mature biofilm can contain multiple species and therefore multiple overlapping conversations. Even within a single species, there may be several quorum-sensing circuits layered on top of one another, each tuned to a different signal and controlling a different suite of genes.

Quorum sensing can regulate virulence factors, secretion systems, enzymes that digest host tissue, and toxins that damage competitors. In biofilms it also influences architecture and maintenance: matrix production, surfactant release, and patterns of detachment. That makes quorum sensing a tempting target. Instead of trying to kill bacteria outright—which often selects strongly for resistance and can fail in the protected interior of a biofilm—one can try to scramble the communication that makes the community effective. This approach is commonly called quorum quenching.

The word *quenching* is borrowed from chemistry and physics, where it means dampening a signal or extinguishing a reaction. In a social metaphor, it is the technology of silence: not destroying the speakers, but making their words meaningless. The appeal is emotional as well as strategic. There is something satisfying about the idea that a fortress can be neutralized by cutting its phone lines.

Two broad strategies dominate quorum quenching. One blocks the receptor—a kind of molecular earplug—so that even if signal molecules accumulate, they cannot trigger a response. The other destroys or modifies the signal molecules themselves, lowering their concentration or turning them into harmless debris. These strategies can sometimes be combined, and both can be built into materials: coatings on medical devices, filters in water systems, or hydrogels applied to wounds.

Receptor blockade is conceptually straightforward. Many quorum-sensing receptors bind their signals with a lock-and-key fit. If a molecule can mimic the key well enough to fit into the lock but not activate it, the receptor becomes occupied but silent. In pharmacology, this is an antagonist. In bacterial communication, it is a jammer.

A famous early family of jammers came from the sea. Certain red algae produce halogenated furanones—small molecules containing a furanone ring decorated with halogens. Their ecological role is often described as antifouling: discouraging bacteria and other organisms from settling on the algal surface. In the late 1990s and early 2000s, researchers realized that synthetic versions of these furanones could interfere with AHL-based quorum sensing in bacteria such as *Pseudomonas aeruginosa*. Work by Michael Givskov and colleagues around 2002 showed that furanone compounds could penetrate biofilm microcolonies and suppress quorum-sensing reporter activity inside them, not merely on the surface. That mattered, because biofilms are notorious for excluding large molecules. Small, hydrophobic jammers can behave differently: they seep into the city.

The results were not always dramatic in the simplest metric of bacterial growth. The bacteria could still multiply. Yet the biofilm's behavior changed: reduced expression of quorum-regulated genes, altered architecture, and in some cases increased detachment and loss of biomass. This was a revealing point for anyone raised on the idea that antimicrobials must kill to matter. Quorum quenching can be more like disarmament than execution. It can reduce damage, disrupt coordination, and make a community easier to clear by other forces: immune cells, physical flushing, or conventional antibiotics.

An anecdote from laboratory practice captures the difference. Students growing *Pseudomonas* in flasks often learn, to their annoyance, that the culture can turn greenish due to pyocyanin and other pigments. Those pigments are not decorative; they are part of a virulence toolkit and are often quorum regulated. In experiments where quorum sensing is disrupted, the color can fade even when the bacteria are still present. The absence of the familiar pigment becomes, in the most literal way, a visible sign of a silenced conversation.

Receptor blockade has limitations. Antagonists can be outcompeted if signal concentrations become very high. Bacteria can mutate receptors to reduce antagonist binding. And, crucially, not all biofilm-relevant behaviors are controlled by quorum sensing in every context. A pathogen in a mouse lung is not a bacterium in a flow cell, and regulatory networks can reroute around a broken circuit. Quorum quenching is therefore not a universal off-switch. It is a way of pushing the system into a less coordinated state, often enough to tip the outcome when combined with other pressures.

The second approach, signal destruction, has a different flavor. If receptor antagonists are earplugs, signal-destroying enzymes are garbage disposal. They remove the message from the environment so that receptors never receive it. The elegance lies in leverage: an enzyme can act catalytically, destroying many signal molecules rather than binding one at a time.

AHL signals are particularly amenable to enzymatic quenching because they contain chemical features that can be attacked. Lactonases can open the lactone ring. Acylases can cleave the amide bond, separating the acyl tail from the homoserine lactone headgroup. Either modification can render the molecule unrecognizable to its receptor. Some bacteria naturally produce these enzymes, which may give them an advantage in mixed communities: they can silence competitors' coordination and keep public goods from being mobilized against them.

What makes this more than a biological curiosity is that enzymes can be engineered. Protein engineering can tune stability, activity, and durability under harsh conditions. That matters because the real world is harsh: disinfectants, temperature swings, drying, and solvents can destroy proteins. A quorum-quenching enzyme that works beautifully in a buffered laboratory solution may fail on the surface of a catheter or in an industrial cooling system. Recent work has focused on making acylases and related enzymes more robust so they can function in coatings and devices, where they must remain active over time.

Signal destruction also changes the evolutionary game. If a population is exposed to an antibiotic, the benefit of resistance is direct and often enormous: survival versus death. If a population is exposed to a quorum-quenching enzyme, the pressure is different. The bacteria are not necessarily dying; they are being forced into a less effective mode. That can reduce the selection for single-step resistance mutations, though it does not eliminate evolution. Bacteria can increase signal production, alter receptors, switch to alternative signaling pathways, or reorganize gene regulation so that crucial traits become less quorum dependent. The hope is that quorum quenching can be, in some contexts, a gentler pressure—less likely to generate explosive resistance—while still providing practical benefit.

The idea of silencing microbial communication has an intellectual ancestry older than the phrase *quorum sensing*. Long before the molecular mechanisms were mapped, microbiologists noticed behaviors that depended on population density: bioluminescence in marine bacteria, pigment production, sporulation, and secretion of extracellular enzymes. In the 1970s and 1980s, the study of luminous *Vibrio* species helped crystallize the concept that bacteria could regulate gene expression based on an accumulating signal. That work, along with later discoveries in pathogens, reshaped microbiology. It forced an uncomfortable admission: bacteria are not merely solitary chemical machines. They are social in a way that matters.

The leap from understanding to sabotage is recent, and it has been driven by the stubbornness of biofilm-associated infections. Chronic wounds, implanted medical devices, and the lungs of people with cystic fibrosis have all served as brutal reminders that antibiotics often behave differently in biofilms. Part of the difference is physical: poor penetration, binding to matrix, and gradients. Part is physiological: slower growth, stress responses, and metabolic states that reduce susceptibility. Quorum sensing intersects with this landscape by shaping behaviors that intensify the problem—protective factors, coordinated virulence, and sometimes the maintenance of the matrix itself.

A concrete experimental demonstration of therapeutic plausibility came when synthetic furanones were tested in animal infection models. In mouse lung infections with *Pseudomonas aeruginosa*, quorum-sensing inhibition reduced the expression of quorum-regulated virulence factors and improved clearance and survival outcomes in some studies in the early 2000s. The details vary with compound, dosing, and model, and no one should pretend that a mouse lung is a human lung. But the conceptual point stood: silencing can matter in living tissue, not only in glass.

Quorum quenching also makes intuitive sense as a partner to the dispersal mechanisms described earlier. Dispersal is a conversion from resident to traveler. Communication is one of the ways residents coordinate their defenses and maintain their infrastructure. If a biofilm's cohesion relies partly on continued agreement—a shared decision to invest in matrix and public goods—then disrupting the conversation can make the city easier to fracture. In some bacteria, quorum sensing influences surfactant production that affects detachment. In others, it modulates enzymes and factors that shape the matrix. The relationship is not universal, but it is common enough that the two themes often meet: chemical silence can be an indirect demolition.

Yet there is a moral complication, even for a reader who prefers science without moral language: silencing a pathogen may feel safer than killing it, but it can have unintended consequences. If quorum sensing restrains certain behaviors in some contexts, blocking it could release them. If a biofilm is composed of multiple species, silencing one conversation could change the balance in ways that favor another, perhaps worse, organism. And because quorum quenching does not necessarily reduce cell numbers immediately, it can be tempting to overinterpret early success. A treatment that makes bacteria quieter may make symptoms better without fully eliminating the infection.

That ambiguity becomes clearer when one remembers what quorum sensing is actually measuring. It is not a mystical "group mind." It is diffusion and confinement. In a well-mixed flask, signal molecules spread evenly. In a biofilm, signals can be trapped locally, creating microdomains of intense communication. A quorum-quenching enzyme on a surface might erase signals near that surface but leave communication intact deeper in the biofilm. Conversely, a small antagonist might diffuse widely, but its concentration could vary in the maze of channels and dense matrix. The success of a blackout is therefore often a story of physics as much as chemistry: diffusion, binding, flow, and geometry.

This is why materials science keeps appearing in the quorum-quenching literature. Imagine a catheter whose surface slowly releases a receptor antagonist. Or a wound dressing impregnated with a stable lactonase. Or a filtration membrane coated with an enzyme that degrades AHLs as water passes through. These are not merely delivery systems; they are spatial strategies. They place the quenching where communication is most consequential: at the interface where bacteria attach and where the first microcolonies begin to coordinate. The goal is to prevent the moment when a few cells become a community.

Even in the best case, quorum quenching rarely acts alone. A blacked-out biofilm may be less virulent yet still physically present. It may produce less matrix yet still adhere. It may become more fragile yet not spontaneously vanish. The most realistic vision is not that silence will cause a city to evaporate, but that silence will make the city governable: easier for the immune system to patrol, easier for antibiotics to finish, easier for fluid flow to wash away.

There is a final, almost philosophical fascination to quorum quenching. The modern world often responds to microbial problems with brute force: stronger disinfectants, broader antibiotics, higher doses. Quorum quenching proposes a different posture. It treats microbes as responsive agents embedded in environments. Instead of escalating firepower, it alters incentives and information. The bacteria remain alive, but they are less certain of their numbers, less coordinated in their spending, less capable of synchronized aggression.

A blackout does not need to be total to be useful. In a crowded city, even partial loss of communication—a broken radio tower, jammed frequencies, garbled messages—can be enough to slow emergency response, disrupt supply lines, and fracture collective action. For bacteria, whose coordination can hinge on a handful of molecules per cell, a modest reduction in effective signal can shift the balance from cooperation toward confused individualism. In that confused state, a biofilm's celebrated advantages begin to look less like inevitabilities and more like temporary achievements, vulnerable to a quiet kind of interference.

## 9.3 Engineered Sieges

When a city is built to withstand assault, a single battering ram rarely wins. Biofilms have had a long time to practice this kind of resilience. They slow the entry of hostile chemicals, they shelter slow-growing cells that shrug off antibiotics aimed at rapid metabolism, and they recruit a chemistry of their own—enzymes that neutralize threats, pumps that expel toxins, and shared public goods that buffer stress. The result is a familiar human experience: a surface is cleaned and seems safe, and then a week later the same stubborn film returns; a wound is treated and improves, and then inflames again; a catheter works and then clogs; a patient feels better on antibiotics and then relapses.

The temptation is to escalate the same tools. Stronger antibiotic. Higher dose. Longer course. More aggressive disinfectant. Those approaches sometimes work, and sometimes they fail in ways that feel almost mocking. The reason is not only resistance in the usual genetic sense. Biofilms can be tolerant even when they are not resistant: the drug can be "right" and the bacteria can still survive because the biofilm's structure and physiology change the rules of engagement.

Engineered sieges are an attempt to change those rules deliberately. Instead of asking a single molecule to do everything—penetrate the matrix, reach the cells, enter the cell, hit the target, and kill—these strategies split the problem into fronts. Some attack the inhabitants. Others attack the infrastructure. Some make the surface itself less hospitable, so the city never gains a foothold. The best of them borrow tricks from biology's own long war: predators, enzymes, and physical design.

Bacteriophages, or simply phages, are the most vivid example of using a biological weapon against a biological city. A phage is a virus that infects bacteria. Under an electron microscope, many phages look like tiny lunar landers: a geometric head packed with genetic material, a tail, and leg-like fibers that bind to bacterial surfaces. Unlike antibiotics, which are usually blunt in their targeting, phages can be exquisitely specific. A phage that kills one strain may ignore its close relatives. That specificity is both an advantage and a complication. It can spare beneficial microbes, but it can also make therapy fragile if the target population is diverse.

Phages have a history intertwined with the history of antibiotics, and the tale reads like an alternate timeline of modern medicine. In the early twentieth century, Felix d'Hérelle championed phage therapy and used it in some of the first deliberate attempts to treat bacterial disease with viruses. Interest surged and then faded in much of the West after antibiotics arrived, not because phages were necessarily ineffective, but because antibiotics were easier to mass-produce, standardize, and prescribe. Phage therapy persisted in different forms in parts of Eastern Europe and the former Soviet sphere. Today, as antibiotic resistance and biofilm-associated infections have become more urgent, phages have returned to the global conversation, now with the added force of genomics, synthetic biology, and better regulatory science.

For biofilms, the phage story has a particular appeal. Phages replicate *in situ*. If they can reach their host cells and infect them, they make more copies of themselves, which can then infect neighboring cells. A chemical drug is finite; a phage is potentially self-amplifying. Yet biofilms impose the same barriers on phages as on many antimicrobials, and some additional ones. The matrix can slow phage diffusion. Binding sites can be hidden. Cells deep in the biofilm may be metabolically dormant, and some phages depend on active cellular machinery for efficient replication. Biofilms can also shed phage-resistant mutants. In other words, phages are not magic bullets. They are living siege engines with their own vulnerabilities.

This is where engineering enters, and it enters in two different senses. One sense is selecting and formulating phages intelligently: cocktails rather than single phages, combinations tuned to the likely diversity of an infection, and dosing strategies that anticipate resistance. The other sense is literally engineering phages to do more than infect and burst cells. In the most imaginative designs, a phage is modified to carry genes that undermine the biofilm's infrastructure. If the matrix is a protective wall, the phage arrives not only with weapons but with sappers.

The simplest version of that idea is to arm phages with matrix-degrading enzymes. Many bacteria produce enzymes that cut polysaccharides, proteins, or extracellular DNA. Some phages already encode depolymerases, enzymes that help them penetrate capsules or biofilm-like layers. Synthetic biology makes it possible to broaden this approach: a phage can be engineered to express an enzyme that degrades a key matrix component, weakening the biofilm as phage infection spreads. The elegance is not only in increased killing, but in improved access. As the matrix loosens, more phage particles and more antibiotics can penetrate. The biofilm becomes less a fortress and more a leaking sponge.

Even without genetic modification, some phages naturally do part of this job. When phage-infected bacteria lyse, they release intracellular contents that can change the local chemistry and weaken nearby cells. The process can create voids and channels in a biofilm, altering its internal structure. In the best case, those changes make the remaining bacteria more vulnerable to the immune system and to antimicrobials. In the worst case, they release clumps that can spread if there is no effective clearance. The hazard is the same one that shadows all dispersal-focused strategies: demolition without containment can turn a localized problem into a distributed one.

A second family of engineered sieges uses enzymes not as weapons carried by viruses, but as standalone tools or as cargo on nanoparticles. If phages are the cavalry, enzymes are the bolt cutters. Biofilm matrices vary between species and environments, but one component appears with remarkable frequency: extracellular DNA (eDNA). This DNA is not simply the debris of dead cells. In many biofilms, it is a structural element that helps bind cells together and anchors other matrix components. Removing it can reduce viscosity and mechanical strength.

DNase, an enzyme that digests DNA, has therefore been explored as an antibiofilm adjunct. The challenge is delivery. Enzymes can be degraded, diluted, or washed away. They can also fail to reach the dense interior. Nanoparticles offer one solution. If DNase is attached to the surface of a nanoparticle, the particle can act as both a delivery vehicle and a local concentration platform: it carries active enzyme into the matrix and degrades eDNA as it moves. Some experimental systems have shown that DNase-coated nanoparticles can reduce biofilm mass and improve antibiotic performance. The point is not that nanoparticles are inherently glamorous, but that they give physical form to a strategy: bring the tool to the wall, not the wall to the tool.

Nanoparticles also allow combinations. A single particle can be designed to carry an antibiotic and a matrix-degrading enzyme, producing a one-two effect in a microscopic radius. The enzyme opens space; the drug follows into newly accessible niches. If that sounds like military language, it is because the biofilm itself behaves like a defended structure. The difference is that the "battlefield" might be a lung, a wound, a urinary catheter, or a food-processing line, and the collateral damage must be considered with unusual care.

Enzymatic strategies are not limited to DNA. Many biofilms rely on specific polysaccharides as scaffolding. One such polysaccharide, poly-$N$-acetylglucosamine (often abbreviated PNAG, and known by related names in different bacteria), appears in a range of pathogens. An enzyme called dispersin B, first characterized from *Aggregatibacter actinomycetemcomitans*, can degrade PNAG. In laboratory studies, dispersin B can detach established biofilms and sensitize bacteria to stresses that otherwise fail to dislodge them. It is an appealing illustration of the "infrastructure" approach: instead of trying to poison the inhabitants directly, take away the protective architecture that turns a cluster of bacteria into a hard-to-eradicate community.

The catch is that matrices are not uniform. Even within a single species, different strains can rely on different materials, and the same strain can build a different matrix depending on nutrients, oxygen, and shear. An enzyme that works beautifully on one biofilm may be underwhelming on another. This is not a reason to abandon enzymatic approaches; it is a reason to think diagnostically. Successful sieges depend on knowing what the wall is made of. In medicine, that is a difficult demand, because clinicians rarely have time to characterize a biofilm's matrix in detail. In industry, the demand may be more feasible, because systems are more standardized and sampling is easier. That difference partly explains why some antibiofilm strategies have found earlier adoption in non-medical settings.

There is another engineered siege that operates even earlier: prevent settlement. Biofilms begin with adhesion, and adhesion begins with a surface. If a surface can be made less welcoming, the city may never be built.

Nature offers a compelling prototype in shark skin. Sharks swim in microbe-rich water, yet their skin is not easily fouled. Part of that resistance comes from chemistry, but part comes from microtopography: fine ridges and patterns that make it difficult for organisms to settle and form stable attachments. Engineers have borrowed this idea to create patterned materials that reduce biofouling. The goal is not to sterilize a surface, which is often unrealistic, but to make attachment mechanically unstable. If cells cannot find a comfortable foothold, they are more likely to be removed by flow or to remain in a transient state that is easier to disinfect.

Micro-patterned surfaces can work by several mechanisms. They reduce the available flat area for contact. They create local curvature that interferes with the alignment of pili and flagella-driven attachment. They alter how water films and proteins coat the surface, which in turn affects adhesion. In some designs, the patterns are on the scale of micrometers, comparable to the size of a bacterial cell, making the geometry directly relevant to how a bacterium experiences the surface.

The most interesting aspect is that these surfaces can be passive. They do not rely on releasing chemicals into the environment. That can be an advantage where toxicity is a concern, such as in implants, or where chemical release would be diluted rapidly, such as in high-flow systems. Passive designs also exert different evolutionary pressures. It is harder for bacteria to evolve around a physical landscape than around a single molecular target, though bacteria are resourceful and can adapt by changing adhesins, producing more matrix, or selecting for variants that attach more effectively. Still, the timescale and the path to adaptation can be less straightforward than with drug resistance.

Of course, a surface that prevents attachment in one context may fail in another. Protein layers from blood or tissue fluids can coat a device, masking microtopography. In industrial systems, minerals and organic films can deposit and change surface properties. The "surface" a bacterium attaches to is often not the underlying material but the conditioning film that forms within minutes. This means that engineered surfaces must be evaluated in realistic conditions, not only in clean laboratory water. Yet even when the effect is partial, it can matter. A small reduction in initial adhesion can delay biofilm development long enough for routine cleaning to be effective, or for immune clearance to prevent establishment.

Some of the most ambitious siege strategies combine chemical cues, physical disruption, and killing in tightly designed sequences. Small molecules that manipulate bacterial signaling networks are one example. Because intracellular messengers such as c-di-GMP help govern the transition between biofilm and planktonic lifestyles, compounds that lower c-di-GMP can push cells toward dispersal and away from matrix production. In principle, such compounds can be used as "softening agents" that weaken the biofilm's commitment to cohesion. The hazard is the familiar one: dispersal without eradication can spread the problem. In clinical language, any dispersal-inducing adjunct must be paired with a bactericidal therapy or an effective clearance mechanism.

This pairing is not merely theoretical; it echoes the logic of many successful combination therapies in infectious disease. One agent makes the pathogen vulnerable; another does the killing. The novelty in biofilms is that vulnerability can be structural. The target is not only an enzyme inside the bacterium but also the architecture outside it.

A vivid way to grasp the promise and the peril is to think of an infected implant. An orthopedic screw, a heart valve, or a catheter is not just a foreign object; it is a landscape. Bacteria that attach can hide from immune cells that cannot easily patrol the surface, and they can persist despite antibiotics that struggle to penetrate the matrix. If a therapy could degrade the matrix and expose the bacteria, immune clearance and antibiotics might succeed without removing the device. That would be a genuine change in patient experience: fewer surgeries, fewer prolonged infections, fewer cycles of relapse. Yet if the same therapy releases a shower of bacteria into the bloodstream without killing them, it could transform a localized device infection into a systemic one. This is why engineered sieges are often designed with multiple redundancies: matrix degradation coupled to bactericidal action, phage replication coupled to enzyme delivery, surface design coupled to controlled release of antimicrobials.

The human story behind these strategies is not only one of cleverness but of frustration. Biofilms are among the reasons chronic infections can be so demoralizing, for patients and clinicians alike. A person with a non-healing wound may endure repeated rounds of treatment that seem to help and then fail. A patient with cystic fibrosis may face infections that cannot be fully cleared despite aggressive care. A hospital may invest in cleaning protocols only to find that biofilms persist in hidden niches. These experiences give emotional weight to what might otherwise sound like technical tinkering.

There is also a quiet shift in mindset. Traditional antimicrobial thinking often assumes that microbes are best addressed by stronger poisons. Engineered sieges treat the biofilm as an ecosystem and a structure. They borrow from ecology (predators and competitors), from chemistry (enzymes and signaling molecules), and from engineering (microtopography and materials). The result is not a single invention but a family of approaches that share a conviction: a biofilm can be made to fail, not only by killing its cells, but by making its way of life impossible.

In the end, the most persuasive image is not a miracle cure but an operational one. A phage particle binds, injects its genetic payload, and turns a bacterium into a factory that will eventually burst. An enzyme clips the strands of DNA that act as structural reinforcement. A patterned surface denies the first settlers a stable grip. Each intervention is modest on its own scale, almost microscopic in its violence, but together they can change what a biofilm is allowed to be: less cohesive, less coordinated, less sheltered, and more exposed to the ordinary forces that clear microbes from healthy systems.

A city that cannot repair its walls, cannot coordinate its defenses, and cannot lay foundations on bare ground does not need to be conquered by a single decisive blow. It empties. It fragments. It becomes, for its inhabitants, an unreliable place to live.

www.ingramcontent.com/pod-product-compliance
Lightning Source LLC
LaVergne TN
LVHW020233210926
848481LV00016B/1847
*9798896652755*